Frank Lowber James

Elementary Microscopical Technology

A Manual for Students of Microscopy

Frank Lowber James

Elementary Microscopical Technology
A Manual for Students of Microscopy

ISBN/EAN: 9783743687257

Printed in Europe, USA, Canada, Australia, Japan

Cover: Foto ©berggeist007 / pixelio.de

More available books at **www.hansebooks.com**

ELEMENTARY MICROSCOPICAL TECHNOLOGY.

A MANUAL FOR STUDENTS OF MICROSCOPY.

IN THREE PARTS.

PART I.

THE TECHNICAL HISTORY OF A SLIDE

FROM THE CRUDE MATERIALS TO THE FINISHED MOUNT.

BY

FRANK L. JAMES, Ph. D., M. D.

President St. Louis Society of Microscopists, Member of American Society of Microscopists, Ed. St. Louis Medical and Surgical Journal.

ST. LOUIS, MO.:
St. Louis Medical and Surgical Journal Company,
1887.

Preface.

IN all existing text-books of microscopical technology with which I am acquainted, not only in the English, but in the French, German and Italian tongues, the technology proper—or the manipulations and processes incident to the preparation of material for microscopical examination, is so mixed up with the micrography—histological, pathological or biological, that it is an almost hopeless task for the student, especially the beginner, working without a master, to separate them. In the multitude of details and the interruptions to continuity caused by the attempt at teaching simultaneously such widely divergent subjects, the elementary student fails to grasp a clear idea of either the one or the other.

Another difficulty, incident to and inseparable from instruction attempted to be conveyed in this manner, is that the author must, at many points anticipate details or presuppose some acquaintance with the subject on the part of the student.

Realizing these difficulties, first as a student remote from skilled teachers, slowly working out by experiment each problem as it arose, and afterwards as a teacher, searching for a text-book to put into the hands of my students, I finally undertook the preparation of a manual modelled after an ideal in which nothing should be taken for granted, no previous acquaintance, on the part of the student, with the subject-matter presupposed, and in which each step of the work, each process and manipulation should be explained in orderly sequence.

This little manual is the result of this idea; how well I have succeeded in its execution is for the reader to say. I will only add that I have endeavored to make it strictly practical, having embodied in it the actual experiences of many years of almost incessant labor in this direction.

The present volume is devoted entirely to Elementary Technology, and details the Technical History of a Slide from the crude materials up to the finished mount. It constitutes Part I of a work on General Microscopical Technology constructed upon the same scheme and plan, the other parts of which will appear in due time.

No. 615 Locust St.,   FRANK L. JAMES.
  St. Louis, Mo.

ELEMENTARY MICROSCOPICAL TECHNOLOGY.

I.

Microscopical technology is a description of those processes and appliances by means of which objects are prepared for examination under the microscope and permanently preserved for future reference and study. In microscopy, as in every other pursuit which involves the use of tools or instruments, the number and nature of the apparatus and appliances deemed necessary for the performance of good work will vary very greatly according to the taste, ingenuity, and above all, the pecuniary ability of the individual. While some men will be content and do good work with a few simple instruments, costing but a few dollars, others will require the most elaborate outfit, costing as many hundreds. In the present series of articles I shall describe only those instruments which experience has taught me to be absolutely essential, leaving the student to learn from more elaborate text-books and the catalogues of the instrument makers, those more complicated appliances which, while they are frequently of great assistance, are not absolutely essential in doing good work and hence must be considered as *articles de luxe*.

§ I. The processes through which an object passes from its crude or natural condition up to the finished slide, ready for the cabinet, vary according to the nature of the material. They may be grouped under six general headings, viz;
   a. Preserving in the mass.
   b. Hardening those substances which are too soft, and softening those which are too hard to be cut with the section knife.
   c. Embedding.
   d. Section cutting.
   e. Staining.
   f. Mounting on slips.
  Some objects have to pass through all of these processes,

while others have to undergo only a portion of the manipulations indicated.

§. II. The instruments and apparatus required for the performance of each step will be described under its proper heading. There are, however, certain tools which are required in almost every stage of our progress, and it will be well for the student to provide himself with them at the outset. They are as follows:

a. A half dozen needles mounted in wooden handles. Four of these should be straight and the others curved. Surgeons needles are the best for this purpose, though ordinary sewing needles will answer for the straight ones. The handles should be five or six inches long and large enough to give a good 'grip' to the fingers. Tapering cedar penholders are excellent for the purpose. The needles should be inserted into the small end and carried deep enough to be perfectly firm. They should not project more than three-quarters of an inch from the handle, as otherwise they are too springy for delicate dissections.

b. Small forceps, straight and curved. The ordinary iris forceps are excellent for this purpose. If dentated, the teeth should be ground or filed out and the ends of the blades brought to a point.

c. Small scissors—straight and curved.

d. A few delicate scalpels. Old tenotomes, or cornea-knives, well sharpened are the very things. Almost every oculist has some such knives which have become useless for his work but which answer admirably for microscopical dissections.

e. A dozen watch crystals. The old fashioned 'bulls-eye' cystals are the best. They may be obtained from any large jewelry establishment for a trifle. Two or three "individual salts" will also be found useful.

f. Three flat saucers, like those used for shirred eggs.

g. A wash-bottle holding 10 or 12 fluid ounces of distilled water.

h. A half dozen medicine droppers or pipettes.

i. A half dozen camel's-hair pencils, assorted.

j. Small spirit lamp and stand.

k. Salt mouths for preserving crude material. Sulphate of morphia bottles fitted with good soft corks are very useful for containing small bits of material in process of hardening, sections, etc.

§. III. The following chemicals are needed at the outset; Alcohol, 95° and absolute; glycerin, pure; benzol, pure, and commercial; oil of turpentine and benzin, of each one pound. The two latter are used for cleansing purposes, as explained hereafter. Chloroform and sulphuric ether, of each 6 or 8 fluid ounces. Picric acid and bichromate of potassium, of each an ounce. Distilled water.

Other chemicals will be described when needed and under the proper headings.

§. IV. Since all organic matter is prone to decay when left to the influences of the atmosphere, our first care must be to properly preserve the material which we are hereafter to prepare for examination. This is best done by means of certain preserving fluids which are modified according to the substances upon which they are required to act. These fluids may be divided into three classes, viz:

a. Those which simply arrest decay.

b. Those which harden or soften as well as preserve.

c. Those which, in addition to other functions, stain the material.

In the first class we have solutions and combinations of the various antiseptics, such as common salt; boracic, carbolic and salicylic acids, chloral hydrate, glycerin, corrosive sublimate, etc.

In the second class we have alcohol, Mueller's fluid, solutions of the mineral acids etc.

The third class embraces picric, chromic and osmic acids, chloride of gold, etc.

Each of these agents has its special use, which must be developed as we proceed in the study of microscopy, since it is manifest that the treatment which is proper for delicate tissues (fœtal tissues, larvæ of insects, desmids or algae, for instance) would not suit for a tumor, a piece of muscle or objects of like description. There are two fluids however which are of such universal use and applicability that to a certain extent they supply the places of all the balance—viz; alcohol and glycerin. It is recommended that the elementary student confine himself as much as possible to the use of these fluids in preserving material for his future work—always bearing in mind the following rule, viz;

Fresh organic matter should never be put into strong alcohol, but treated as follows;

Provide three saltmouths holding a pint each, and having well ground glass stoppers or closely fitting velvet corks. In lieu of these, preserve-jars will answer. They should have a mouth large enough to receive the object intended for preservation without squeezing or forcing it in. In the first of these jars put a mixture of 1 part of alcohol and 2 parts of distilled or rain water. In the second reverse the proportions, making the mixture two parts of alcohol to one of water, and in the third put strong alcohol (95°). The material (which we are supposing to be animal matter, is placed first in jar No. I where it should be left for a period varying according to the size of the object, from 24 to 48 hours. It is then removed to the second jar and left an equal length of time, and finally transferred to the third jar. In the course of time the fluid in the first jar will be so weakened by the fluids abstracted from the material placed in it that it must be thrown away. Jar No. 2 will also have become weakened, but it can now be made to serve for the first bath, and No. 3, for the same reason will act as No. 2. The third jar will thus be the only one that it is necessary to replace from time to time.

Small objects immersed in glycerin will keep for a long time without any special precautions. Delicate objects containing much aqueous matter should not be put directly into pure anhydrous glycerine for the same reason that they should not go at once into strong alcohol—the affinity of the preserving fluid causes it to abstract water from the object too rapidly, thus shriveling and injuring delicate tissues.

Where a number of specimens are put into the same jar at the same time, each one should be marked in such a manner that it may be known subsequently. Colored or white threads may be attached to it and knotted in such a way as to serve as a distinguishing marks. Where the student has plenty of laboratory room and other conveniences those precautions are not necessary, as each object may be kept in a separate receiver.

§. V. The following are formulæ for those preserving fluids most commonly in use. We shall hereafter refer to them only by the title here given:

GOADBY'S FLUID.—Chloride of sodium 2 ounces, alum 1 ounce, corrosive sublimate 1 grain, distilled water 1 pint. For tender tissues add a grain of sublimate and double the quantity of water. For substances containing carbonate of calcium omit the alum and use double the quantity of common salt.

MUELLER'S FLUID.—Bichromate of potassium 25 grains; sulphate of sodium 10 grains; distilled water 2 ounces.

WICKERSHEIMER'S FLUID, as modified by Hager.—Salicylic acid 4 drams, boracic acid 5 drams, carbonate of potassium 1 dram, mix and dissolve, by the aid of heat, in a mixture consisting of 12 ounces of water and 5 ounces of glycerin. When solution is complete add 12 ounces of alcohol at 95° in which 3 drams, each, of oil of cinnamon and oil of cloves have been previously dissolved.

MEYER'S FLUIDS.—Make a stock fluid consisting of 1 part of salicylic acid dissolved in 100 parts of pyroligneous acid C. P

For larvæ, hydræ and nematoidæ, take 3 parts of the fluid and add 10 parts c. p. glycerin and 20 parts distilled water.

For infusoria; glycerin 2 parts, distilled water 8 parts, mix, and add 1 part of the stock fluid.

For algæ, desmids etc., take 1 part of the stock fluid, 1 part glycerin, and 20 parts of water.

In addition to the above the following are the proportions usually observed in making hardening and preserving fluids.

Chromic acid solution—1 per cent.
Picric acid solution—Saturated.
Osmic acid solution—One-half of one per cent.
Chloride of gold solution—One-half of one per cent.
Bichromate of patassium solution—2 per cent.

A TYPICAL MOUNT.

II.

In order to make plain the plan and scope of the present work, and that the student may understand the rationale of the processes through which an object must be carried in its transition from crude material to the finished slide, we will in this present chapter briefly describe the making of a suppositious mount. Each step in the manipulations will be described in detail in the chapters which follow, and will be taken up as nearly as possible in the order in which it comes in the actual progress of the work

§ VI. We will suppose the object to be a pathological specimen, a tumor for instance, recently removed. It is plain that the direct examination of such an object can only be made with very low magnifying powers, such for instance, as may be obtained by the use of a pocket magnifyer or Coddington lens. In order to reach the histological elements we must use high powers, and these can only be used by transmitted light—that is light sent through the object to be examined. We must therefore contrive a method by which the object, or a *representative portion* of it, may be made translucent. This may be done by taking a small portion of the material and mashing it out very thin between two pieces of glass. In former times this method was frequently resorted to, but as it could manifestly yield but very distorted results it has long since been abandoned by those who use the microscope as an instrument of precision. The other alternative at our disposal is the cutting of a section from the object with a very keen knife; and here we meet with another difficulty viz.; the object is (usually) too soft to offer such resistance to the passage of the blade as will enable us to cut a section of sufficient and uniform thinness. It is true that formerly such sections were cut with a Valentine's knife and which were supposed to be thin enough to yield practical results, but the device is now very rarely resorted to. The object must therefore be submitted to a process which will harden it and at the same time preserve it. If we are in a great hurry to arrive at results we may attain the desired end by freezing our object, but we will suppose that in the present instance resort is had to one of the hardening and preserving fluids already mentioned or yet to be described. Having hardened the material, our next step is to cut it into thin sections. Formerly this was accomplished by holding the hardened object in the hand and slicing off a section with a razor. This process is no longer used in exact and scientific work. Free-hand cutting has given place to the microtome or section cutter—just as in exact work free-hand drawing has yielded to photography. The object is therefore transferred to a section cutter to be sliced into sections, and as these sections must be made extremely thin and uniform, it must be securely held in the microtome. It must be arranged so that it can be fed to the knife and at the same time have no lost lateral motion. This necessitates embedding it in some liquid material that will harden around it and hold it firmly in place. This done, the section knife is brought into play and the object is sliced to the requisite degree of thinness. Here we must digress a little in order to explain subsequent operations.

If we take any very thin substance—say a piece of paper, and place it under the microscope in a dry state, we will find on examination that we get a very insufficient idea of its intimate structure. If we moisten the object with water we find that many details of the structure are brought out and shown us which were invisible under the former examination. If instead of water, we use glycerin or Canada balsam, the structure is rendered still more distinct, and if the specimen be only thin enough the minutest detail is thus finally brought into view.

Let us suppose, further, that this bit of paper consists of two or three, or more, kinds of fibres—say silk, cotton and linen, all of the same color and so interwoven with each other that it is impossible for the eye to follow the ramifications of either material. It is plain that if we can find a dye or stain which will attack the cotton and not the silk or linen, or *vice versa*; or that stains cotton one shade or hue, silk another, and linen another, the problem of differentiating the elements which enter into the structure is a very simple one.

With these two hints as to the reason why the sections are put through the next two processes, and leaving the philosophy of the same to be explained hereafter, we will resume the progress of our slide toward completion.

The sections as they fall from the knife are received in a vessel filled with fluid—water, glycerin or alcohol, according to circumstances, and when a sufficient number has been cut we pick out one of the thinnest and best, and place it in the staining fluid, where we will leave it while we prepare the glass slip upon which it is to be mounted.

We take for this purpose a piece of clear glass 3 inches long and 1 inch wide, the edges of which have been ground and polished, and placing it on an instrument called a turn-table, with a pencil dipped in cement, (the nature of which depends upon the fluid which we shall use as a mounting medium), we spin a ring in the center of it. This ring is large enough and deep enough to receive the object to be mounted, and should be allowed to get quite dry before the slip is used. The object, in the mean time, has been removed from the staining fluid and put through a number of little details to fix the stain, clear away the embedding material, etc., and is now soaking in the fluid which is to serve as a mounting medium, the functions of which are to render the object transparent and preserve it against decay. In this instance we will suppose that glycerin has been chosen as a medium. The ringed slip, thoroughly cleaned, is now placed on the mounting box (a frame with a glass top and provided with a mirror so arranged as to throw the light upwards through the top and objects laid on it) and a drop of pure glycerin is allowed to fall in the center of the ring; the object is quickly transferred from the glycerin bath in which it has been lying, placed on the drop of glycerin already on the slip, and arranged in the position it is henceforth to occupy; air bubbles are gotten rid of as hereafter explained, the cover glass is applied, and clamped in position; surplus glycerin is washed away and the slip and cover glass carefully and thoroughly dried with prepared blotting paper. The clamp holding the cover glass in place is now removed and the slip transferred back to the turn-table where a ring of cement is spun around the edges of the cover glass. This ring is allowed to dry, and the slip is again washed and carefully dried before a second layer of cement is applied. The final touches which vary according to the taste, skill etc., of operator are then given to the slide; it is labelled and if the job is properly done, is good for an indefinite number of years.

§ VII. Such, in brief, is the ordinary routine of processes usually employed in making a mount of a pathological or histological specimen of the soft tissues. There are many minor operations, matters of detail, entirely omitted or barely alluded to in the foregoing sketch, while many of the manipulations there described are varied according to the nature of the mounting medium finally chosen; but an analysis of the processes enables us to divide them into three principle groups, as follows, viz;
1. Those pertaining to the preparation of the material.
2. The preparation of the slide to receive the object.
3. The mounting of the object on the slide, including finishing.

§VIII. The operations belonging to the first group include all of those enumerated in § I except the last (f). We will repeat them, subdividing them as we go along, as follows;
1. Preservation of the material in mass.
2. Hardening soft structures and softening hard ones.
3. Embedding. The details of this series of processes vary greatly, according to the nature of the structure of the material (i. e. solid matter, such as muscle, requiring a different treatment from vacuolar or spongy matter, like lung tissue), the kind of microtome used, etc. They may be roughly subdivided into prepartory manipulations, (such as freeing from the preserving or hardening fluid, dessication of the surface of solid matter, dipping into a fluid which prevents adhesion of embedding material, dehydration of spongy tissues, soaking in liquid embedding material to fill areoles etc.), and the actual operation of embedding which consists of a series of several manipulations which we need not enumerate.
4. Section cutting—including the treatment of the sections, preparatory to staining, such as freeing from embedding material, etc.
5. Staining, including the fixation of stains, dehydration, clearing with essential oils, etc., and saturating with the fluid which is to be used in the final mounting medium.

§. IX. The second group embraces the following manipulations and processes
1. Cleaning the slips, and cover glasses.
2. Preparation of cements, etc.
3. Making cells, the use of the turntable, etc.

§X. The third group of operations bring together the finished results of the other two. In it the finished section is placed upon the prepared slide. It consists of
1. The arrangement of the section or object within the cell.
2. Filling the cell with the mounting medium.
3. Removal of air bubbles and foreign matter.
4. Putting on the cover glass and arranging it in place.
5. Washing off surplus mounting medium, and drying the slide.
6. Returning the slide to the turntable, centering, etc.
7. Final closing of the cell by cementing the cover glass to the cell wall.

In the chapters which are to follow each process will be taken up in the order in which it occurs in this schedule or scheme of the technology of mounting, and with the explanations given in this chapter before him the student will be enabled to follow understandingly each step that is taken.

III.

§ X. HARDENING.—Most of the preserving agents mentioned in our opening chapter, have also hardening powers, the method of action being about the same in all of them, the variation being in the degree of hardness attained and the length of time required to reach the requisite point. Thus alcohol is by far the best and most useful of the hardening agents for most animal tissues, but when applied to the softer tissues of the lower organisms, by withdrawing the aqueous matter of which they are largely composed, it shrivels and shrinks them so that they are practically worthless for subsequent observation and study. Osmic acid, which in weak solution is a perfect hardening agent for the latter class of tissues, is too slow in its action for the former. No absolute rule can therefore be given as to the choice of a hardening agent. In the great majority of cases, however, alcohol, used as already suggested (i. e. in gradually increased strength) will answer every requirement. When a more rapid effect than that produced by absolute alcohol is desired, the latter may be supplemented by picric acid which, however, stains as well as hardens. The same may be said of chromic acid, bichromate of potassium and their combinations (as in Mueller's fluid). The materials and methods used in hardening being so diverse and numerous, we can, at this stage in our work, give only the philosophy and outline of the process, leaving for future explanation the suggestion as to the proper medium to be used for various tissues.

§ XI. The hardening agents usually resorted to in animal histological researches are:
 a. Heat.
 b. Cold.
 c. Dessication.

d. Chemicals, viz.: alcohol, chromic, picric, and osmic acids, etc.

HEAT—Boiling: There are certain tissues the structure of which is rendered visible by submitting them to the action of heat. Such, for instance, are the branched muscular fibers of the frog's tongue, the fibers of the crystalline lens, etc. The object to be examined is put into a large test tube, covered with distilled water and boiled for a few moments over the flame of an alcohol lamp, or in a water bath. This method was formerly much in vogue, especially among English microscopists, but it is now rarely resorted to. If however, one is pressed for time and desires merely to differentiate certain elements in the structure of material such as that referred to, it affords a quick and ready means of so doing—the word differentiate being used in a limited sense, as no one, even a novice, can believe that the true histological features of any structure can remain unaltered after being submitted to an agent or process so destructive to life and living matter.

COLD—Freezing: It is but a very few years since German histologists and investigators maintained that material which had undergone the processes of hardening by heat or chemicals, gave results, when examined under the microscope, that were altogether unreliable and misleading. These teachers advocated the process of hardening by freezing and declared that frozen sections of fresh material were the only ones in which the true histological elements and conditions could be observed and studied. Of late, however, those who advocated the freezing process most strenuously, have found that it produces abnormalities quite as numerous and as serious as does hardening by chemicals, and that it has little to offer in return for the many advantages of which it deprives us. There are, however, circumstances under which it may be advantageously resorted to, and most of the modern microtomes (section cutters) are provided with freezing devices. In the absence of special contrivances the material may be frozen sufficiently hard for all practical purposes by immersing it (contained in a test tube) in any of the well known freezing mixtures, of which the following are the best:

1. Snow or powdered ice, 12 parts; nitrate of ammonium and chloride of sodium, each, 5 parts. This will lower the

temperature from any point to 25° below zero, F.

2. Snow or powdered ice, 4 parts; crystalized chloride of calcium, 6 parts. This will cause a depression of the temperature 82° F. from any point above zero (the actual depression below this point being not quite so much, though very nearly so.)

3. Equal parts of nitrate of ammonium, carbonate of sodium and water. This lowers the temperature 57° F. from any point.

Dr. Beale suggests a method of freezing which is very convenient for dentists and others having nitrous oxide at command. Tissues may be instantly frozen by imbedding them in albumen or gelatine and placing them in a test tube, submiting it to the action of a jet of that gas. It requires but a few moments to render the mass as solid as rock. After freezing the material, very thin sections may be cut from it and afterward prepared and examined at leisure. The advantages to hardening by freezing are that perfectly fresh material may be used and much time thus saved when time is an element to be considered.

DESSICATION. Skin, mucous membrane, and material of this kind may be hardened by simply drying, by exposure to the atmosphere at ordinary temperatures, or by the careful application of a very low degree of heat. To quote Dr. Beale again (whose book "How to work with the Microscope" is one of the very best text-books ever published), tissues that we wish to harden in this manner may be stretched on boards, fastened in place with pins and thus allowed to dry. Sections cut from material thus prepared when placed in water or diluted glycerin, resume, to a great extent, their fresh appearance. When an air-pump is convenient a most rapid and excellent method of dessication is to suspend the article in the receiver over a vessel containing anhydrous sulphuric acid. On exhausting the air the moisture is rapidly withdrawn from the object and taken up by the acid.

CHEMICAL MEDIA. By far the most useful and generally applicable of these is ALCOHOL, used as described in Chapter I, for preserving purposes. The range of animal textures to which this agent is applicable is limited only by the fact that it renders albuminous tissues, otherwise transparent, opaque and granular, and that it precipitates albumen from solutions. Its avidity for water and consequent shriveling action on delicate tissues has already been noted.

ANHYDROUS GLYCERIN will harden many tissues very nicely, but its action is comparatively slow, and to obtain a satisfactory result the glycerin must be kept in an air-tight vessel and changed frequently.

The brain, spinal cord and such tissues are hardened very beautifully by immersion in solutions of CHROMIC ACID, the chromate and bichromate of potassium, and mixtures containing these chemicals, such as Mueller's and other solutions.

PICRIC ACID is another hardening agent of a wide range of usefulness. It is used in either alcoholic or aqueous solution or as an ingredient in complex hardening fluids, such as Kleinenberg's. If used in water the solution should be saturated, since it is but sparingly soluble in that medium. For the same reason the bulk of the solution should be large in proportion to that of the material. When dissolved in alcohol the percentage may be varied according to the rapidity of the desired effect.

OSMIC ACID. To those whose studies lie among the lower forms of animal and vegetable life this chemical is one of the most valuable in the whole list of hardening agents. Its application not only kills those organisms but it does instantly and without distortion. It preserves the tissues indefinitely, and by its selective staining property (which will hereafter be alluded to in its proper place) it brings out features that would otherwise escape observation. It should be used in solution not stronger than one per cent., and this solution should be kept absolutely protected from the action of the light. I use for this purpose a glass stoppered bottle, thickly painted with drop black and afterward covered with heavy tin or lead foil. The whole should be placed in a closely covered metallic box (a yeast powder can is the very thing.) In preparing this solution the following precautions should be observed: Clean the bottle which is to hold it, in the most perfect manner, rinsing frequently with distilled water. Clean also in the same manner the outside of the glass tube in which the osmic acid is found in commerce. Weigh the distilled water or measure it carefully, pour it into the prepared bottle, and then drop the tube of osmic acid into it. Put in the stopper and shake the vessel sharply until the glass tube is fractured and the water allowed to reach the acid. Even when thus prepared the solution is prone to decompose and must be made afresh quite frequently. Specimens to be hardened in this agent should be cut up quite small—a rule by the way, that holds good with all other agents, and put into a vessel prepared as suggested above for the exclusion of light. The period required for hardening varies according to the object.

SOFTENING AND DECALCIFICATION.

IV.

§. XII. Hardening Vegetable Tissues.—The remarks on hardening hitherto made are supposed to apply to tissues of animal origin only, but there are certain vegetable tissues which are also too soft and friable to stand cutting while in their natural condition. In such cases where it is not desired to preserve the shape and relative positions of the parts, the objects may be hardened by simple drying. As this process almost invariably shrivels and distorts a specimen it is better to resort to other and less objectionable methods.

Here again, alcohol is the best medium with which we are acquainted, and it may be used exactly in the manner described for hardening objects of animal origin. Sometimes, however, the specimens have colors which it is desirable to retain and which would be extracted or changed by this hardening medium. When such is the case the ordinary processes of hardening must be avoided, and resort had to a device which answers the purpose by mechanical methods. This process, which consists of saturating the object with some inert material which melts at a low heat but is solid at ordinary temperatures, will be described under the head of embedding.

§. XIII. Softening.—The process of softening tissues to fit them for the section knife is a very simple one and is applied, as a general thing, only to the stems and roots of plants, seeds with a hard coat or shell, etc. The articles to be treated are placed in rain or distilled water and left to macerate for several days. In summer time the normal temperature of the atmosphere will be sufficient, but in cold weather the process is hastened by placing the vessel containing the material close to the fire or in some warm place. If the object be very dense (a bit of cocoanut shell, or vegetable ivory, for instance) a gentle artificial heat may be applied.

After macerating from 24 to 48 hours, the objects, if not sufficiently soft, should be transferred to dilute alcohol (50°) or methylated spirit of the same strength, where they should be left for a similar length of time. If on trial they are still found to be too hard to cut, the process should be repeated from the beginning. There are very few vegetable substances which will not yeild to this treatment.

Roots or stems containing gummy or resinous matter insoluble in water or dilute alcohol, should be soaked in fluids in which their resins are soluble. Benzin is a good general solvent for gums and resins, and is much cheaper than benzol, chloroform, ether and other fluids usually recommended for this purpose. Where benzin is inefficacious the other solvents should be tried.

The lower forms of vegetable life, the fungi and parasitic plants and the tender shoots of the higher forms, require but little preparation for the knife.

§. XIV. DECALCIFICATION. Certain forms of organic mattei of both animal and vegetable origin consist of soft tissues covered by, or permeated with a harder material of mineral or earthy nature. The greater part of this hard material, in both animals and plants, consists of the carbonate or phosphate of lime, or both these salts, (though some plants carry a large amount of silex). It is frequently desirable to rid such material of its mineral constituents in order that the softer parts may be intimately examined. When such is the case recourse is had to the process known as decalcification. The agent used for this purpose is a very dilute solution of one of the mineral acids, either nitric or muriatic. In his work on the microscope the late Dr. Carpenter gives the following directions for the employment of these erodent materials:

"When the lime is in the state of a carbonate, as in the skeleton of the ecchinoderms, the body to be decalcified should be placed in a glass jar or wide mouthed bottle holding from 4 to 5 ounces of water, and the acid added gradually, drop by drop, until the disengagement of bubbles shows that it is taking effect. The solvent process should be allowed to take place very gradually, more acid being added as required. When, on the other hand, much of the lime is in the shape of phosphate, as in bones and teeth, the strength of the acid solvent must be increased. For the hardening of the softer parts of the organic matrix (after being freed from lime) chromic acid should be used. In the case of small bones, such as the cochlea of the ear, a one-half of one per cent solution of chromic acid will not only dissolve the lime but at the same time harden the matrix. Larger masses of bones require the addition of nitric or hydrochloric acid to the chromic, the proportion being 2 per cent nitric or 5 per cent of hydrochloric. By some investigators the acids are mixed at the beginning of the process, but others recommend that the bone should lie in the chromic acid solution for a week or ten days before the second acid is added. If the softening be not complete in a month, more acid must be added. When thoroughly decalcified the matrix must be transferred to alcohol and kept in it until needed for section cutting."

§. XV. Much discretion must necessarily be used in the selection of the proper agents and methods to be employed in preparing material for the knife—as indeed in every other department of microscopical technology, and the student without a teacher will frequently be at a loss how to proceed. He must rely upon himself, and must expect to make many failures and mistakes; but by following the ways and means suggested by us, he will eventually succeed, and will be all the better posted by the experience gained in overcoming difficulties. Experience, after all, is the great teacher, and the knowledge that is to guide one in doubtful cases is rarely to be obtained from text-books and manuals, no matter how elaborate, practical and complete they be. They can only point the way, but individual experiment alone can make the successful worker.

V

§ XVI. EMBEDDING.—Our material having passed through the various operations of hardening, softening, or decalcification, as the case may be, its next step in its progress toward a completed slide, is its preparation for the section cutter or microtome, as the instrument in which it is sliced into thin sections is technically called.

The various forms of microtomes will be described in the ensuing chapter, with illustrations showing the best patterns for histological work. At present it is sufficient to say that in its simplest form the section cutter consists of a receptacle for holding the material to be cut, a screw or other apparatus for feeding it to the knife, and a razor or knife with a very keen edge. Formerly the razor alone was employed for this purpose, the material being held in the left hand and the cutting instrument used free handed. This method is still adhered to by some European microscopists, but as it bears the same relation, in regard to exact work, as the photograph does to free sketching, or the old fashioned two-foot rule does to a micrometer scale, it has no place in modern exact science.

Whatever form of microtome be used, some method must be devised for holding the material to be cut, so that it shall be properly supported on all sides and that it shall have no motion, lateral or vertical, save that given to it by the feeding device. Where the object is dense and the sections not required to be of any great degree of thinness, this may be secured by wedging it into the section cutter by means of some substance itself easily cut without damage to the knife. Formerly, this plan was much resorted to, carrots, potatoes, alder-pith and such like material being used for the wedge.

This method is still available for many operations and it is well for the student to bear it in mind; but for all purposes of exact histological, pathological or biological research of to-day, the method of embedding is by means of a material which is solid under ordinary circumstances but which may be rendered fluid by the application of a low degree of heat or by dissolving it in some rapidly evaporating medium, such as ether, benzol, chloroform, etc. The object is immersed in the fluid and the latter is allowed to harden around it, thus fitting itself to the minutest portion of its exposed surface, and holding it firmly within its embrace.

§ XVII. The choice of an embedding medium depends upon three things, viz;
1. The nature of the object to be cut.
2. The temperature at which the cutting is accomplished.
3. The form of microtome used.

It is plain, for instance, that a very hard substance, giving considerable resistance to the knife, should not be embedded in a very yielding material; or that an embedding material which is sufficiently soft and tenacious at 60° F. might be too hard and brittle for a freezing microtome, or too soft and fluid to be used when the thermometer ranges in the summer heats; or, again, that an embedding material which can be poured into the well of an ordinary microtome (of the 'army medical' pattern for instance), would not do for fixing the object to the cork holder of a machine microtome of the Thoma or Schanze pattern (see illustrations in next number).

For these reasons microscopists have devised quite a number of ingenious embedding mixtures suitable to every class of work, the best of which only need be described here. In this, as in every other department of an elementary work, the multiplication of formulæ serves merely to confound the student with an *embarras de richesse*.

§ XVIII. For the general run of histological or pathological work undertaken by physicians, and for use in the form of section cutter most commonly employed, there is no embedding material with which I am acquainted that is any better than PARAFFIN—used either in its natural state, or mingled with other fatty or oleaginous matter. By the judicious selection of the various grades of paraffin, or mingling any one grade of it, as a basis, with solid fats or oils, as the case may be, an embedding mixture may be made which will suit almost any required emergency.

§ XIX. The materials most generally used as adjuncts or corrigerents of paraffin, are cacao butter, lard, wax, spermaceti, vaselin, tallow, suet, and olive or refined cotton-seed oil. For ordinary work, in a room where the temperature averages 65° or 70° F. during the winter months, I have found a mixture of equal parts of paraffin (hard), cacao butter and mutton suet, to work very well, especially if a few drops of sweet oil be added to the stock occasionally, after frequent meltings and remeltings. Prof. Stowell, of the University of Michigan, in his "Manual of Histology" (an excellent text-book, by the way, for those who desire to pursue this line of microscopical study) gives the following formulæ for the embedding mixtures used in his laboratory:

Soft. Solid paraffin, three parts; cacao butter, one part; lard, three parts.

Hard. Solid paraffin, three parts; cacao butter, two parts; spermaceti, one part.

Harder. Solid paraffin, two parts; cacao butter and spermaceti, of each, one part.

For a very soft and transparent medium, easy to cut, he recommends a mixture consisting of two parts of paraffin, and one part of vaselin.

Among other solids which may take the place of paraffin, none is better than a vegetable wax, known in commerce as bayberry tallow (or myrtle wax, as it is called in the Southern States, where it grows in great plenty). It may be used alone or in mixtures, as in the case of paraffin. It is brittle at low temperatures and hence had best be cut with a little olive oil, if the weather be very cold. It is insoluble in cold alcohol but readily soluble in that medium at a temperature of from 110° to 130° F. This latter quality makes it quite valuable as an embedding material for certain kinds of work, of which, however, more will be said when special technology is taken up.

Albumin, from hen's eggs, is also a valuable embedding medium. It is best when used, as suggested by Runge, in combination with tallow, according to the following formula;

Take fresh white-of-egg and, after removing the chalazæ, add to each one hundred parts of albumin, fifteen parts of the officinal liquor sodii, and shake them together in a large test tube.

When well mixed add forty parts of melted beef tallow and again agitate until a homogeneous mass is obtained. One advantage of this mixture is that it is unaffected by alcohol, and an object once embedded in it, the entire mass may be removed from the section cutter and placed in that fluid for indefinite keeping.

In the appendix or supplementary chapters of this series, other embedding materials will be described in detail. At present we must content ourselves with those given above, which the student using the ordinary microtomes will find amply sufficient for all practical purposes.

§ XX. PREPARATION OF EMBEDDING MIXTURES.—The ingredients should be melted together in a porcelain or glazed earthen-ware vessel, provided with a cover fitting sufficiently closely to prevent the ingress of dust and foreign matter. The melting should be done in a water-bath, as otherwise there is danger of burning or scorching the fats. If heated directly over a lamp scorching may be avoided, however, by watchfulness and constant stirring. Enough of the material to last for many months may be made up at one time, and when required for use a portion sufficient for the work in hand may be withdrawn and remelted. The surplus material removed from the microtome may be used over and over again, if the precaution is taken of straining it occasionally. Even this precaution may be avoided if the remelting is done with a gentle heat and the vessel containing the liquid material be allowed to stand in hot water until all foreign matter settles to the bottom.

§ XXI. METHODS OF USING.—The preparatory treatment of material to be embedded varies according to its nature and structure—matter which is sufficiently dense throughout to need no internal support (a bit of liver or of muscle, for instance) requiring a different treatment from one which is areolated or spongy (as a bit of lung tissue).

1. In the first instance the material is removed from the preserving or hardening fluid, the adherent surplus of which is allowed to drain back into the bottle, and it is then placed on a piece of heavy bibulous paper or of porous tile (a bit of a broken flower pot is excellent for this purpose) and allowed to remain there until the surface is quite free from moisture. This exter-

nal dessication may be hastened by frequently turning the material, or by the application of soft bibulous paper or cloth, which application should be made with as little pressure as possible, in order to avoid distortion of the relations of the histological or anatomical elements. Care must also be taken not to allow the drying process to proceed too far, as otherwise the material may become distorted by shriveling. When the surface is sufficiently dry the object is taken on the point of a dissecting needle and dipped into a weak solution of gum arabic or gelatin, withdrawn and again allowed to dry. This process should be repeated a second or even a third time, the object being to give a coating of a substance easily gotten rid of by washing in water, and which will prevent the adherence of the embedding material to the edges of the sections.

I wish to impress upon my readers the importance of this preliminary, though it will require but a single trial to convince one of its utility as a time saver. When it is neglected the sections must frequently—in fact, almost invariably, be placed in some fluid solvent of the embedding material, which itself is often very hard to get rid of, and which sometimes injures delicate structures submitted to its action.

While the coating of gum or gelatin is drying the embedding material may be put into the water-bath and melted. As soon as it becomes entirely fluid the source of heat should be withdrawn, as it is not desirable to heat it much beyond the melting point. The object still remaining on the point of the needle, is dipped into the melted material two or three times, the coating received each time being allowed to harden before a second is given. This ensures its complete envelopment at all points. with the embedding matter. The well of the microtome, being clean and freed from dust, is now filled with the fluid embedding material, the object is lowered into it and arranged in the desired position and held there, still on the point of the needle, until the material sets around it, when the needle should be withdrawn. The well should be filled quite full and the surplus displaced by the object allowed to run over on the surface of the microtome. The material, shrinking in cooling, would otherwise leave a depression in the center, thus failing to support a part of the object. For the same reason it is well to warm the microtome somewhat before using it, as when the process of cooling is too rapid the material shrinks away from the side of the well, thus occasioning lateral motion in the whole mass and preventing very accurate section cutting. The method of avoiding the results of such an accident will be further treated of in the succeeding chapter.

§ XXII. EMBEDDING AREOLATED OR SPONGY TISSUE —When it becomes necessary to cut material which is full of minute cavities, like a bit of sponge, lung, or other areolated tissue, the process of embedding must be modified somewhat; since not only the external surface, but much of the internal structure, needs support. In such cases we proceed as follows:

The object is removed from the hardening or preserving fluid, freed of its excess and dried exactly as in the foregoing cases. If it be dipped in the gum-arabic solution, all excess of this material must also be squeezed out most carefully. In very delicate tissues which will not stand much manipulation it is, therefore, best to dispense with the preliminary dipping in gum-arabic water, and proceed to place the object directly into some fluid embedding material which will enter into the areoles and there harden. This embedding material may either be identical with that which it is proposed to use for the external support, or it may essentially differ from it. In most cases the same material will answer for both external and internal support; but sometimes, from the nature of the internal structure, this is not advisable. A most excellent material for filling the areoles of delicate tissues is gelatin melted in a very small amount of water, and with the addition of a few drops of glycerin to each ounce of the fluid. Gelatin, when used for this purpose, should be put into cold water and allowed to swell for ten or twelve hours at a temperature of not over 75° F. The excess of water is then carefully drained off and the vessel containing the gelatin submerged in the water-bath until melted. While still hot add 15 to 20 minims of glycerin to each ounce of the liquid, stir until homogeneous and remove from the water-bath. When the solution has cooled to a point at which it will no longer injure the delicate tissues, but is still fluid, the object may be immersed in it and left there until the solution jellies. A knife or spoon is then passed around it and it is lifted from the surrounding gelatin, the excess of which, adherent to the outside, must be carefully removed before placing it into an oleaginous embedding mixture. The latter must be of such composition that it will remain fluid at a comparatively low temperature, as otherwise the gelatin filling the areoles will again become fluid and some of it escape.

For the great majority of areolated tissues it will suffice to place them into the melted embedding material, after it has somewhat cooled off but before it begins to solidify, and leave them there until they become saturated. Where even this amount of heat will be hurtful the filling must be effected by immersion in a saturated solution of paraffin in oil of turpentine. Where resort is had to this device the material should be placed first in absolute alcohol, and thence transferred to oil of turpentine, before its final immersion in the paraffin solution. More will be said on this subject when special technology is taken up.

§ XXIII. EMBEDDING FOR MACHINE MICROTOMES.—In this class of microtomes (See plate I, Fig. 3) the object to be cut is fastened to a holder, usually of cork, the embedding material having been previously poured around it or applied afterward, layer by layer. Formerly the paraffin and other mixtures already given, were used for this purpose, and, indeed, are still adhered to by a great many investigators,—the object, after preliminary preparation, being placed in a small box made of writing paper and the embedding material poured around it. Very small objects are easily embedded by laying them on a piece of solid paraffin (cold) in which a cavity is made by touching it with a bit of heated wire. The object is dropped into the cavity and the liquefied paraffin hardens around it almost instantly, leaving it ready for the knife in a very few minutes.

Within the past two years a new embedding material called celloidin has come into favor with users of the machine microtomes, especially with those who cut serial sections (that is, sections each of which bears a definite relation to the one preceding and succeeding it). This material, which is a patented article, made in Germany, is simply pure pyroxylin or gun cotton. It is found in commerce in the shape of shreds which resemble very closely those of gelatin or isinglass. It is soluble in equal parts of sulphuric ether and alcohol (97°) and when used for embedding, the solution should be saturated, to make which requires from 24 to 30 hours digestion, with frequent agitation. The object to be embedded should in the meantime be put to soak in a similar mixture of ether and alcohol and left until thoroughly saturated; or it may be soaked (as advised by Prof. Libby, in a paper read before Sect. G of the American Association, 1884), first in absolute alcohol for one or two hours; next in strong ether for six hours; and lastly, in a weak solution of celloidin in alcohol and ether for an equal length of time. When ready to embed, a good cork, which has previously been soaked in absolute alcohol, is chosen and a small quantity of the saturated solution of celloidin is placed on the end of it, and allowed to dry until it is well stiffened. Then the specimen, with as much of the embedding solution as will adhere to it, is placed on this and allowed to dry partially, when another coating is placed over it. The proceeding should be repeated until the specimen has a sufficient quantity of the embedding material around it to hold it firmly. It is then allowed to dry, and when stiff is placed in alcohol of 80° for at least 12 hours before cutting. (Prof. W. Libby in the paper referred to.)

In the laboratory of the N. Y. College of Physicians and Surgeons the above method is sometimes used, but more generally the proceedure is as follows: "The specimens are embedded in paper boxes, in the usual way; or a cork is wrapped with one or two layers of thick writing paper, allowing it to project an inch or an inch and a half above the surface of the cork, thus making a round box of paper with the cork for a bottom. Into this box pour a small quantity of the embedding fluid and let it dry. The specimen, having been previously soaked, as directed above, is now placed in the box, adjusted to position and allowed to dry for 10 minutes, when the balance of the box is filled with the embedding fluid. The box is exposed to the air until the entire mass has become semi-solid and then immersed in weak alcohol (alcohol 1 part, water 2 parts) where it is left for 24 hours, when it is ready for cutting."

The method of staining and clearing sections of material embedded in this way will be explained under the proper heading.

§XXIV. THE SECTION CUTTER OR MICROTOME. As explained before, the earlier investigators contented themselves with a simple razor or sharp knife and cut their sections free handed. The first attempt at improvement on this method was Valentine's knife, an instrument consisting of two very thin, sharp blades, placed parallel to each other and so arranged that the distance between them can be regulated by a set-screw. This knife is of no service whatever in exact work; and as to the old free-hand method, while one who is skillful may occasionally, or even quite frequently, make sections of small surfaces of great thinness and evenness, it is not to be relied upon in those operations where every section has a definite value (as in serial sections.)

Fig. 1. Army Medical Museum Microtome.

§ XXV. Fig 1 represents the form of microtome most commonly used in this country, and known as the Army Medical Museum pattern. It consists of a bed-plate of brass into one end of which is set a round plate of thick glass perforated in the centre with a hole one inch in diameter. To the under side of the plate is screwed a brass cylinder having an inside diameter exactly the same as that of the hole in the glass surface plate. This cylinder, which is technically termed "the well," is provided with a closely fitting piston and follower, worked by a screw with a milled head, which constitutes the feed apparatus. The object embedded in the well, as described in Chapter V, is moved upward to the knife by a turn of the feed screw, the threads of which are usually 64 to the inch; consequently one full turn would give a vertical movement of $\frac{1}{64}''$, a half turn $\frac{1}{128}''$, a tenth of a turn $\frac{1}{640}''$, etc. The apparatus is provided with a clamping arrangement for attaching it firmly to the work table. When desired this pattern is also furnished with a pan which fits closely under the bed-plate and around the feed cylinder, and which can be filled with a freezing mixture, so that it can be used as a freezing microtome. With a little practice very thin and even sections of small objects can be made with this form of microtome.

A modification of this form of instrument is made by W. H. Bullock, of Chicago. The general principles are the same in both microtomes, though I think that Mr. Bullock's device has the merit of greater convenience and that the feed-screw in his pattern, being supported in the base of the instrument, works more evenly and surely than in the older Army Medical Museum instrument.

For cutting hard substances the same general forms of microtome are used, the only difference being in the fact that the well is provided with a clamp and set-screw, working from the side, the function of which is to hold the object to be cut more firmly than it could be held by embedding mixtures.

A very clever device for use with the Army Medical Microtome and instruments of that description, is shown in Fig. 2. It is the invention of Dr. Seiler of Philadelphia, and is known as his knife carrier. Its method of construction and use is so well shown in the engraving and is so obvious that it needs no description. The knife may be made with a movable handle as shown in the cut, or the arms may be made to carry an ordinary section knife.

§ XXVI. MACHINE MICROTOMES.—Although the idea of an instrument in which the cutting apparatus forms an integral part, is not a new one—microtomes of this description having been made and used a half century ago, it is only within the last five or six years that any great attention has been paid to them, either in Europe or America. The very rapid strides made in other departments of biological research, and especially in the improvement of microscopic objectives, rendered necessary the perfection of instruments by which exceedingly thin and uniform sections could be prepared with certainty and rapidity. The demand was met by the appearance in Europe of the microtomes of Ranvier, Schanze, Thoma

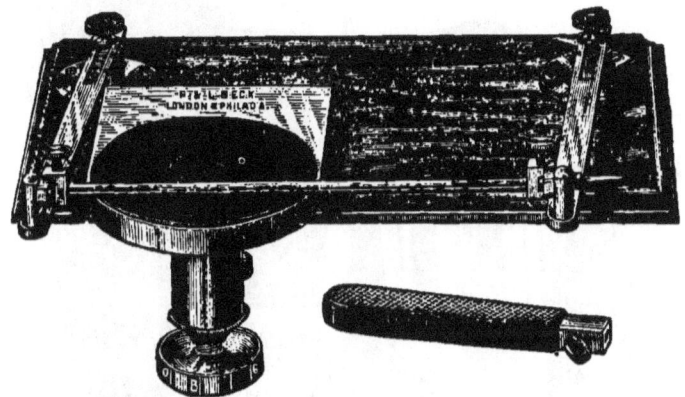

Fig. 2. Seiler's Microtome Attachment (from W. H. Walmsley & Co., Philadelphia.)

and others, several of which were immediately imported into this country by institutes of learning and private individuals. It was not long, however, before American workmanship evolved from the best features of the imported articles, an instrument surpassing them in solidity, lightness, beauty of finish and accuracy of workmanship. Prominent among those who tackled the problem in this country was Mr. W. H. Bullock, of Chicago, W. H. Walmsley, of Philadelphia, and the Bausch & Lomb Optical Company, of Rochester, N. Y. The cutting and feeding parts of all are upon the same general principles and Mr. Bullock has added to his instrument a device for removing the cut sections in series upon an endless band, which is a very great convenience to students of biology and histology.

The plate on the next page represents the Bausch & Lomb large laboratory microtome. The engraving shows the parts of the instrument so clearly that a description of it is

PLATE 1. FIG. 3.—Bausch & Lomb Microtome

scarcely necessary. The base, arm and bed of the knife and object holder are made of a single casting, which insures the maximum of strength and rigidity with the minimum of weight. The knife is fitted with a curved and slotted handle which enables the operator to use any part of the blade, and within the block which carries it is a spring which so regulates the resistance that it passes through objects of varying hardness without deviating from its plane or requiring a change of pressure on the part of the operator. The object holder is attached to a carriage which is adjustable, so that any degree of obliquity may be given to the material to be cut. The feed-screw is provided with an exceedingly delicate and accurate micrometric arrangement which insures absolute accuracy in gauging the thickness of the sections. The same firm makes a smaller instrument which is a simple modification of the one here pictured, for the use of students.

Fig. 4. Walmsley's Microtome.

Fig. 4. represents a modification of Rivet's Microtome, manufactured by Mr. W. H. Walmsley, of Philadelphia. The principle upon which the feeding device is built, as shown in the engraving, is that of the inclined plane; the screw giving great delicacy and accuracy to the movement of the object carrier, thus enabling the operator to secure sections of great and uniform thinness. This microtome is exceedingly simple in construction and very serviceable.

XXVII. THE SECTION KNIFE.—Whatever form of microtome be used, the knife is the most important mechanical factor in successful section cutting; and the more accurate the feeding devices and other accessories, the greater the necessity for a perfect cutting edge. In the machine microtome the knife is essentially a part of the instrument, and hence need not be considered here, save to remark that the blades are usually much larger and heavier than those used with the hand cutting-instruments. As in other matters of detail, there has been much difference of opinion among experts as to the best form of blade, both as to shape and size. Some have recommended a short blade with a bellying edge, like a razor; while others, myself among the number, think that a longer blade is preferable, with an edge perfectly straight from point to heel. Whatever shape be adopted, however, all agree that one side of the blade (that which is to lie next to the surface of the microtome) should be ground perfectly flat, while the other should be hollow-ground, as shown in Fig. 7, which is a section through the blade of the section knife used by myself.

Fig. 5.—Section through blade.

The blade should be from 5″ to 7″ in length, from $\frac{7}{8}$″ to $\frac{17}{16}$″ in width, and from $\frac{3}{16}$″ to $\frac{1}{5}$″ in thickness at the back. The material should be the very best razor steel, tempered to a hardness but little, if any, short of that of the razor. Not one in a dozen of the blades ordinarily in use are properly and homogeneously tempered; and since, as remarked, the beauty and thinness of the sections depend largely upon the evenness and fineness of the cutting edge, this is a most important point to be considered in the selection of a knife. It is almost needless to add that the blade should be kept as keen as hone and strap can make it; and hence a few words on the subject of sharpening will not be inappropriate here—especially as the remarks apply to surgical cutting instruments as well.

THE BEST HONES are the Turkish razor or barber hones, and those made from the finer grades of the Hot Springs or Ouachita stone. The former is the quickest cutter, but the latter makes an edge that is improved but little by strapping. The stone should be sufficiently wide and long to allow the blade to rest entirely upon it. The surface should be kept as true as possib e and in honing a section knife most of the attrition should be upon the hollow-ground side. The reason for this latter precaution will be apparent to anyone on glancing at Fig. 5, and is, namely, that there being much less bearing surface on the hollow side the process of honing progresses much more rapidly and evenly than it could upon the flat side. After getting the blade nearly to an edge the flat side should be passed a few times over the hone, holding it so that every point is in contact with the surface of the stone. The best strap is a piece of soft French calf-skin, thoroughly freed from dust and dirt and tightly stretched on a smooth, flat piece of wood. A knife should never be put upon a hone or a strap until both have been carefully wiped. A single grain of the fine, sharp sand or dust with which the atmosphere is so frequently laden may, by nicking the delicate edge, cost one an hour's hard honing.

Fig. 6.—Section Knife.

THE HANDLE of the section knife should be large enough to afford a good grip to the fingers and should be roughened after the fashion of those of surgical instruments. The section knife should never be used for any purpose other than for cutting sections, and it should never be put away after use until it has been thoroughly cleaned, wiped dry and passed once or twice lightly over the strap. A soft rag or bit of chamois, charged with *putz pomade*, or the finest jeweler's rouge, is most excellent for preserving the polish, not only of knives but of other instruments and apparatus..

§ XXVIII. THE OTHER ACCESSORIES for section cutting, beside the knife and microtome, are as follows;

A small vessel of alcohol for wetting the knife blade;

A saucer for receiving the cut sections;

Two camel's hair pencils, rather large; one of these is for use in the alcohol, and the other for removing the cut sections from the blade;

A pair of delicate forceps;

Alcohol, glycerin and distilled water;

Wash bottle, cloths for wiping the blades, etc.

§. XXIX. ARRANGEMENT OF THE WORK TABLE. The object being embedded, and the microtome being placed in a convenient position (if of the 'Army Medical' pattern it is clamped to the edge of the table or board), we are at last ready to commence section cutting. The little vessel of alcohol (an acetate of morphine bottle answers admirably), with its camel's hair pencil, should be placed in front and slightly to the left of the microtome. The saucer for receiving the sections should be placed conveniently, to the front and right of the microtome. It should be about half-filled with clean distilled water, water and alcohol or even alcohol alone, as the case may be—the fluid varying according to the nature of the material to be cut, and upon the character of the subsequent mount. This is a discretion which is to be gained only by experience; though it will be manifest to anyone, for instance, that a section of a bit of material injected with gelatin carmine, should not go into water or glycerin. The camel's hair pencil for removing the sections from the knife should lie in this saucer.

Near by and within convenient reach, should be a receptacle for the surplus embedding material from time to time removed from the microtome. The instruments likely to be needed during the course of the work in hand should also be placed within easy reach—a rule which applies to all kinds of technical work, since time and annoyance are thus saved, and nothing is such a foe to good, even work as constant interruption from any source whatever. A peice of soft linen or cotton cloth, for wiping the knife after each cut, completes the arrangement of the working apparatus.

§ XXX. CUTTING.—We are now ready for the first cut. We turn up the milled head of the feed screw until the object to be made into sections is brought to a point where it will engage the edge of the blade passed along the surface of the microtome. If much surplus embedding material has run over the surface we remove it with a scalpel or pen-knife and give the glass plate a thorough wiping to remove any adherent dust or grit. This is an operation that must be repeated frequently during the progress of our work. The section knife is then siezed in the right hand and, after being well flooded with alcohol, is placed, flat side down, upon the surface of the microtome and with an even and uniform motion carried entirely through the object and the surrounding embedding mass.

Here is another point in section-cutting upon which experts differ. Some prefer to make the cut *from* the person, outward; while others claim to achieve better results by drawing the knife *toward* the person. While this is entirely a matter of habit and opinion, it has an importance from the fact that in ordering a section knife it is necessary to state which method is pursued; since the knife, which in cutting toward one would have the flat-ground side of the blade next the microtome plate, would present that side upward in cutting from the person. I prefer to cut toward the body, though sometimes—as in cutting very hard substances, the other way is the better, since it enables one to steady the end of the blade with the left hand, which also aids in forcing it through the resistant tissues.

After making the first incision, the object should be examined closely for the purpose of ascertaining whether the embedding material has surrounded it perfectly at all parts. If, as frequently happens, from imperfect dessication or other causes, air bubbles should have formed around it, these cavities must be filled up with freshly melted material. Sometimes also, as mentioned before (C. V. §21), the embedding material shrinks so much in cooling that an appreciable space is left between it and the walls of the well of the microtome, thus giving rise to a slight lateral motion in the entire mass. When this space is very minute it may be remedied by leaving a portion of the embedding material uncut by the section knife after traversing the object, the projection afterward serving as a rest for the forefinger of the left hand, which thus steadies the mass as the knife passes through it. A better plan, however, is to pass a heated knife blade into the embedding material, close enough to the sides to cause sufficient of the melted material to flow into the crevice and fill it up. Any vertical motion of the mass, other than that given it by the feed screw, is prevented by a button attached to the top of the piston.

Having assured ourselves of the absence of faults in the embedding mass, or having remedied them, we next proceed with a scalpel or pen-knife blade to remove, for a limited depth, all embedding material from in front of the object, the presence of which, while lending no support, serves as an obstacle to the edge of the knife and must otherwise be traversed before the latter can engage the object to be cut. This done we are prepared to make the first section.

The blade is again drenched with alcohol and laid flat upon the top of the microtome in such a manner that its heel will be the first part to come in contact with the object. The milled head of the feed screw is manipulated with the fingers of the left hand and the object thus brought upwards through a distance equal to the thickness of the proposed section. The knife is then brought toward the person (or shoved from it, as the case may be), with an even and steady motion, drawing it at the same time so that it traverses the object from the heel toward the point of the blade. Care must be taken to keep the knife flat upon the microtome. This is another most important point in successful section cutting, the failure to preserve the contact at all points being the most fruitful source of failure with beginners in this part of the work.

When, at length, the knife has traversed the object and the section is lying upon the blade, the latter is carried over to the saucer and the point dipped into the fluid therein; the camel's hair pencil, laden with the liquid, is applied to the blade just above the section and the latter is floated off into the receptacle designed to receive it. The knife is wiped, again flooded with alcohol, and the operation of cutting repeated until the object is cut up, or until a sufficient number of sections are made.

The feed screws of many of the microtomes in general use are cut 50 threads to the inch, so that one entire turn of the milled head gives a section of $\frac{1}{50}''$, one fourth of a turn $\frac{1}{200}''$, etc., thus enabling the operator to gauge the thickness of his sections very accurately. The beginner is prone to try to cut his sections very thin and, as a consequence, makes many failures, sometimes spoiling a good deal of valuable material without getting a single passable section. He should make his first attempts on comparatively thick sections, and after he has learned to hold his knife perfectly steady, gradually decrease the thickness of his cuts.

§ XXXI. CARE OF INSTRUMENTS.—When a sufficient number of sections have been obtained, the first care of the operator should be to clean and put away the instruments that he has been using. The knife should be dried and strapped, as described, and the balance of the embedding material removed from the

microtome. If any portion of the object remain and it be
something which one desires to keep, the embedding material is
cut away from around it and the fragment returned to the
preserving fluid from which it was taken. When celloidin or
albumin is used as embedders the entire mass may be dropped
back into alcohol.

The microtome should be thoroughly cleaned and dried before it is put away. Care should be taken, especially, to clean
the feed screw, as a very slight amount of erosion here may destroy the exactness of fit upon which so much of the accuracy and
delicacy of section cutting depends.

§ XXXII. If from any cause it becomes necessary to suspend the operation of section cutting before it is completed, the
object may be hermetically sealed into the embedding mass by
resorting to the method described for filling the space left by
shrinkage, viz.: by passing a heated knife blade into the mass and
smearing a portion of the melted material over the exposed surface
of the object. If this or some similar precaution be neglected, the
object soon dries up and is shriveled and distorted until it no
longer has any histological or anatomical value.

VII.

§XXXIII. Preliminary Treatment of the Sections. The sections are now ready to undergo the varied manipulations through which they must pass prior to the final act of mounting, viz.; staining, dehydration, freeing from still adherent embedding material, etc. If the object which we have been cutting was solid, the embedding material will give us little or no trouble, as it will have separated from the section spontaneously, owing to our having taken the precaution of dipping the object in the gum solution prior to embedding it. If, on the contrary, it was porous, spongy or areolated, like a piece of lung tissue, and we have been compelled to fill its cavities with embedding material, the process is a more tedious one, involving two or more steps, according to the nature of the material, the embedder and the subsequent mount. The first is dehydration, or removal of the water which has been absorbed by the sections. This is done by removing the latter into alcohol of 95°, if they have been lying in water, and thence into absolute alcohol. If they were deposited in dilute or commercial alcohol when they were cut, they should be removed thence directly into absolute alcohol. The proportion of the fluid should be large in comparison with the mass of the cut sections, for reasons that are obvious. From the alcohol they should be removed into one of the essential oils—oil of turpentine, bergamot, etc., or into benzol, to dissolve out the embedding material. If the subsequent mount is to remain unstained and to be made in balsam, the section may be removed from the solvent oil directly into a solution of balsam in benzol, chloroform or turpentine, and there left until ready to be put on the slip. If, however, the mounting medium is to be glycerin, glycerin jelly, or any of the aqueous media, the steps must be retraced exactly in reverse order to that in which they were from alcohol to water or glycerin. But before any of the above described steps be taken there is another process for which the sections are already prepared, as they lie in the saucer of water, or alcohol, and which can be performed better at this than at any subsequent stage of progress, viz., STAINING.

§ XXXIV. As suggested in one of the earlier chapters of this work (Chap. II. § VI.) the process of staining serves a three-fold purpose in microscopical technology, viz.:

1. By giving a tint to very delicate, transparent membranes, tissues, or parts of objects, it enables the eye to detect and follow details which would be otherwise invisible.

2. By the selective properties of some stains, which readily impart color to certain elements, tissues or substances and are inert upon others; or by the comparative readiness with which different tissues, or different elements of the same tissue, absorb the same stain, but with varying degrees of intensity, staining enables a differentiation of elements or parts thereof that would otherwise have the same general appearance under the microscope. Thus, osmic acid stains fat-cells black and protoplasmic matter gray; picro-carmine imparts to certain tissues a yellow or orange, and to others a reddish or carmine hue; applied to epithelium, hæmatoxylon stains the nucleus deeper and quicker than it does the balance of the cell. To differential staining, as applied to bacteria and the lowlier forms of life, we owe some of the latest and most important discoveries in biological and pathological research.

3. In addition to these advantages in differentiation, certain stains facilitate the study of tissues by imparting cool and grateful tints to them, thus enabling the eye to dwell upon them for a long time without fatigue or injury to the organ.

Aside from the practical advantages enumerated, the beauty and general appearance of mounts of most animal and vegetable tissues are greatly enhanced by judicious staining.

§ XXXV. From what has been said it will be plain to the student that the subject of stains and staining is a very important one to the microscopist in whatever branch of the science he may elect to pursue his investigations, and one which, in the light of comparatively recent advances in the study of the germ or parasitical origin of infectious diseases, daily acquires more weight and significance. Of what far-reaching consequence, for instance, was Koch's discovery that the bacilli of tubercle will absorb certain of the aniline colors and cling to them with such tenacity that not even nitric acid will bleach them? The judicious application of stains becomes, therefore, of paramount importance to the worker in micro-botany, no less than to the biologist, the histologist and pathologist. So great a field can be but barely touched in the present work, where we must confine ourselves to general principles, leaving for special treatises the technology of bacterial and similar investigations.*

§ XXXVI. While the processes of staining animal and vegetable tissues for examination under the microscope are not a "mystery past finding out", confined to a few professionals in France and Germany, as was thought in this country but a few years ago; and while everything about them is rational and simple, the art is, nevertheless, one requiring considerable judgement, patience and attention to detail. Exactly what stain should be applied to a given tissue in order to bring out all its hidden details, how long to leave any object in contact with the stain, etc., are points to be learned from experience and practice only, aided by the experience of others in similar cases.

§ XXXVII. From the multitude of formulæ met with in text books on microscopy and the current literature of the day, I have selected those which from my own experience, as a practical microscopist and teacher, I have found the most generally useful and meritorious stains to place in the hands of the student and practitioner. There are many others which in certain departments of the study are useful and even indispensable, which I will not mention at this stage of our progress, reserving them for chapters on special methods. In the arrangement of the stains I have followed the old plan of grouping them into simple or non-selective, and selective, though, as a matter of fact, there are no stains but which may not under proper manipulation, be made to do selective tinting.

---

* For the technology of the bacteria investigation I would recommend the following text books, viz.: Methods of Bacteriological Investigation, by F. Hueppe, N. Y., D. Appleton & Co., and The Technology of the Bacteria Investigation, by Chas. S. Dolley, Boston, Cassino & Co.

§ XXXVIII. THE CARMINES.—The most generally useful of all staining materials at the command of the microscopist, whether he be a worker in animal or vegetable histology, is CARMINE. Its diffusibility, high staining power, brilliancy and persistency, combine to make it a stain of almost universal applicability. Among the best of the carmine fluids are the following, viz.:

1. *Ammonia Carmine* or *Beale's Carmine*: Carmine (No. 40) 20 grains; stronger water of ammonia, 1 dram; alcohol 1 ounce; glycerin and distilled water, of each, 4 ounces. Rub the carmine into a paste with a small portion of the distilled water, being careful to break down all lumps: put into a flask and add the ammonia; agitate for a few moments and then bring to a boil in the water bath, and continue to boil until solution is complete. The flask must now be set aside to cool, leaving the mouth open so that the residual ammonia may escape. Mix the glycerin and the balance of the water and add to the carmine as soon as the latter is cold; agitate, and finally, add the alcohol. Filter and rack off into small bottles with well ground glass stoppers. Properly prepared this solution will keep a n indefinite period without thickening or precipitation. If, however, it should show any signs of separation this process may be corrected by the addition of a few drops of ammonia. This stain may be used either full strength or diluted, as explained hereafter.

2. *Borax Carmine*. Put 60 grains of carmine in a porcelain capsule; pour over it 2 drams of stronger water of ammonia and triturate well until rubbed smooth. Set aside for 24 hours, or until the free ammoniacal gas has escaped. Make a saturated solution of borax in distilled water and of this add, very gradually and with constant rubbing, 4 ounces to the carmine in the capsule. Pour into a bottle of sufficient size and agitate until every particle of carmine is dissolved. Let stand for a few hours, filter and rack off into small glass-stoppered vials.

3. *Woodward's Carmine*, or *Violet Carmine*. To one part of borax carmine, made as in No. 2 above, add 4 parts of distilled water and 8 parts of alcohol of 95°. A crystaline precipitate will be thrown down, which may be separated on the filter or by decantation; wash in alcohol of 95° and redissolve in distilled water (4 parts of water to 1 part of the carmine crystals). This solution is of a bright lilac color, which is changed by the fixing solution (see below) into roses or red.

4. *Thiersch's Carmine.* Carmine, 60 grains; stronger ammonia water, 1 dram; distilled water, 3 drams: mix. In a separate vessel dissolve 24 grains of oxalic acid in 1 ounce of distilled water, and add to it 12 drams of alcohol of 95°. Mix the two solutions. This gives a dark red stain that is exceedingly permanent, and which requires no fixation.

5. *Czokor's Carmine.* Dissolve in 6 ounces of distilled water, 8 grains of burnt alum, and in the solution triturate thoroughly 50 grains of cochineal. Boil for a half hour in the water bath; remove, filter and add distilled water sufficient to bring the solution up to 6 ounces. This stain is a deep cherry red and possesses many advantages as a staining agent, chief of which is the fact that tissues which have been hardened in chromic acid, or potassium bichromate solution, take it as well as do those hardened in plain alcohol.

FIXING SOLUTIONS.—All of the carmine stains, with the exception of Thiersch's, require to be 'fixed' or 'set', and several agents have been suggested for this purpose. Dr. Beale recommends a 1 per cent. solution of acetic acid in alcohol or glycerin. For fixing Woodward's carmine and borax carmine, a solution of hydrochloric acid in alcohol (acid 1 part, alcohol 7 parts) is recommended. When this fixer is used it must be watched very closely, as it bleaches rapidly and destroys very thin sections in a short time. I am in the habit of using for all carmine stains a fixer consisting of 24 grains of oxalic acid dissolved in 1 ounce of distilled water. It acts remarkably well with the stronger ammoniacal solutions.

The method of using the 'fixer' will be explained hereafter.

§ XXXIX. HÆMATOXYLON, OR LOG-WOOD STAINS.—These are also very useful to the student of histology, whether employed in plain or selective staining. Log-wood has the property of staining the nuclei more deeply and rapidly than the remainder of the cell material. It requires no fixing or setting, but as usually made, it has the great disadvantage of not being permanent, either as an applied stain or in solution.

1. *Aqueous Solution of Log-wood.* Dissolve 1 part of alum in 8 parts of distilled water and add to the solution a sufficient quantity of the aqueous extract of log-wood to give the requisite tint, or a deep violet color. (Frey.)

2. *Another and Better Aqueous Solution* is made by the following process: rub together in a mortar 1 part of the ordinary extract of logwood and 3 parts of alum. When reduced to a very fine powder and thorougly mixed add a small quantity of water and continue the trituration, gradually adding more water, until the solution, when examined by transmitted light assumes a dark violet hue. The amount of water depends upon the freshness, etc., of the extract used. When the requisite color is obtained, to each ounce of the solution add two drams of alcohol of 75°.

3. *Alcoholic Log-wood Stain.* While the aqueous extracts of log-wood are prone to decompose, even when the utmost care has been expended in preparing them, the following alcoholic solution will be found to improve with age: Make a saturated solution of calcium chloride in alcohol of 70°, and another of alum in alcohol of the same strength. Mix the two solutions in the proportion of 1 part of the former to 7 parts of the latter. Make a saturated solution of extract of log-wood in absolute alcohol and add it, drop by drop, to the mixture, until the latter assumes a deep purple color. Filter and preserve in glass stoppered vials.

VII.

XL. SELECTIVE STAINS.—While the stains enumerated and described in the foregoing chapter do not by any means embrace all of the simple, or what are usually termed non-selective, tinting solutions, those there given are sufficient for all practical purposes. The student of elementary histology and the practitioner will rarely or never need any others. We may, however, mention that all, or nearly all, of the aniline colors may be used for this purpose; but as they are usually employed in double or multiple staining we have relegated them to a place under this heading.

Of the selective stains, properly so called, none is so generally useful to workers in every branch of microscopy as picro-carminate of ammonium or as it is usually termed, picro-carmine, which is prepared as follows:

1. *Picro-carmine crystals:*—With the aid of heat dissolve in the smallest possible amount of the stronger water of ammonia, 15 grains of best carmine (No. 40) and to the solution add sufficient distilled water to make one fluid ounce. In another vessel dissolve 75 grains of picric acid in boiling water sufficient to make, when cold, a saturated solution. When cool mix the two solutions, put into a bottle with a well ground glass stopper, agitate vigorously and set aside to stand (with an occasional shaking) for several days. At the expiration of four or five days all of the uncombined picric acid will have subsided to the bottom, and the clear supernatant fluid may be decanted or filtered off. The filtrate should now be poured into shallow flat dishes, loosely covered to keep out the dust, etc., set aside in a moderately warm place, and allowed to evaporate spontaneously. The process may be hastened by the very careful application of a low heat, on the water bath, but any attempt at rapid evaporation will injure the brilliancy of the product. The result of evapora-

tion will be a crop of minute, brick-dust red crystals, 10 grains of which, dissolved in an ounce of distilled water, makes the picro-carmine solution, so useful to students of microscopy. Since more or less dust must get into the solution in process of preparation, it should always be filtered before final bottling, and the filtrate should be kept in well ground glass stoppered vials.

An *alcoholic picro-carmine solution* may be made extemporaneously, as follows: dissolve 2 grains of carmine in 30 minims of stronger ammonia water and add sufficient distilled water to make one ounce. Dissolve 8 grains of picric acid in 1 ounce of alcohol and mix the two solutions. After filtration the stain is ready for use. [*The Microscope*, September, 1884.

2. *Indigo Carmine*, or sulph-indigotate of sodium solution.—Take of the best indigo, in lump, 30 grains, powder in a large capsule and dry thoroughly in a water bath. When dry add drop by drop, 2 drams of fuming sulphuric acid (the Nordhausen acid) stirring constantly with a glass rod. The indigo swells enormously under this treatment, and hence the necessity for a large capsule. The whole of the acid being added, cover closely and let stand for 24 hours. At the end of this time add 3 ounces of distilled water, transfer the mixture to a tall flask, and let stand for 4 days, giving the flask an occasional shake in the interim. A magnificent blue color is now attained, but its acidity is so great that it can not be used in this shape, and therefore a strong solution of sodium carbonate must be added. This addition must be made very cautiously, a little at a time and with frequent testing, as an excess of the alkali causes the indigo to separate in a doughy mass. Filter the neutralized solution and evaporate to dryness in a water bath, with a low heat. The resultant powder is sulph-indigotate of sodium. To make the staining solution, dissolve 1 part of the powder in 50 parts of distilled water.

3. *Chloride of Gold.*—Dissolve 10 grains of gold chloride in 1 ounce of distilled water. Keep in a glass-stoppered, dark bottle, covered with tin foil to keep out the light.

4. *Nitrate of Silver.*—This solution should be made of the strength of 5 grains to the ounce of distilled water. If both ingredients be chemically pure the light has no effect on the solution. It should be kept in glass-stoppered bottles.

5. *Osmic Acid.*—This chemical comes in little sealed glass tubes, each containing 1 gram (15½ grains). It possesses a very

disagreeable and powerfully irritant odor, and is at the same time exceedingly sensitive, when in solution, to the chemical action of light. Great care must therefore be exercised in preparing this reagent. To prepare a one-per-cent solution choose a 4 ounce vial of dark glass, with a well-fitting, ground-glass stopper. Make this thoroughly and chemically clean, by washing it first with hot liquor potassæ and rinsing with warm water. Wash a second time with nitro-muriatic acid and rinse, first w tli warm water, then with alcohol, and lastly with distilled water. The stopper, which should also be similarly cleansed, is then put in and the bottle closely enveloped in heavy tin or lead foil, which should extend up to the very rim of the mouth, and under the bottom. The sealed glass tube containing the acid should next be made chemically clean externally, by a similar process. When this is done, weigh into the bottle 1543 grains of freshly distilled, chemically pure, water; sieze the cleaned tube with a pair of clean wooden forceps and drop it into the bottle, replace the stopper, and then, by a violent and sudden motion of the bottle, break the tube. The fragments of glass, being clean, may be left in the bottle, which should be kept in a baking powder box, or some similar case with a closely fitting cover.

These details may seem needlessly exacting; but the fact is that the solution of osmic acid is so prone to decomposition that even with these precautions it can rarely be kept for any length of time. As the chemical is costly ($2.25 to $2.50 per tube of 1 gram) it becomes a serious tax upon one who desires to make frequent use of its most valuable properties. *

§XLI. THE ANILIN COLORS.—These stains have of late acquired an importance in microscopical research second to that of no class of reagents in its entire technology. Their value has been especially shown in bacteriological investigations, and in the domain of biology. Those principally used in histological research are as follows:

1 *Magenta* (or rose-anilin) : solution, 1 grain to the ounce of alcohol (95°).

2 *Blues.*—There are several of these, the most useful being (a) methyl blue, in alcohol to saturation: (b) *acetate of mauvein*.

---

* For the uses to which osmic acid may be advantageously appled in histological research I would refer the reader to the files of the ST. LOUIS MEDICAL AND SURGICAL JOURNAL for 1885—more especially to page 320 (Osmic acid in histology).

4 grains to the ounce of alcohol: *(c) Nicholson's blue*, 1-6 grain; nitric acid 2 minims; alcohol, 1 ounce.

3 *Greens.*—There are also several of these that are useful, viz: *(a) iodine green*, 1 grain to 6 drams of distilled water: *(b) methyl green*, 5 grains to 1 ounce of distilled water: *(c)* a beautiful green is made by mixing the saturated solution of methyl blue (above) with a saturated solution of picric acid in alcohol.

4 *Eosine.*—4 grains to the ounce of distilled water and alcohol in equal parts.

§XLII. The student of histology will find the foregoing list of stains amply sufficient for all ordinary purposes. Any more at the present time would simply serve to embarrass and confuse him. For the convenience of the reader I have grouped the stains into a tabular diagram by means of which he may see at a glance the function of each stain, and especially those stains which are supplementary to each other.

General Stains.
- Picric Acid,
- Ammonia Carmine,
- Eosine.

Selective stains
- Beale's carmine,
- Borax carmine,
- Woodward's carmine,
- Logwood,
- Indigo-carmine,
- Anilin blue,
  } Simple stains, not requiring the action of light or reagents
- Gold chloride,
- Silver nitrate,
- Osmic acid,
  } Simple; but requiring the action of light.
- Picro carmine,
- Carmine and indigo carmine,
- Logwood and anilin blue,
- Logwood and gold chloride,
- Logwood and silver nitrate,
- Gold chloride and silver nitrate,
- Methyl blue and picric acid.
  } Double Stains.

Stain in bulk and harden at the same time.
- Osmic acid,
- Picric acid,
- Gold chloride,
- Borax carmine and alcohol,
- Logwood and alcohol,
- Eosine and alcohol,
- Chromic acid, or bichromate of potassium.

[This table is a modification of one published by J. W. Groves in the American Journal of Microscopy, Feb., 1880.]

§XLIII. APPARATUS ETC.—The apparatus for staining are few and simple and nothing is required beyond those things already enumerated, viz: an assortment of watch-glasses, some pipettes for transferring stains (each stain should have its own pipette), a wash bottle full of distilled water, a couple of deep saucers with camel's hair pencils, and a pair of small wooden forceps, fill the list. A few goblets or wine glasses to cover the watch-glasses are always in order.

§XLIV. THE APPLICATION OF THE STAINS.—The older histologists held in high esteem the processes for staining in the mass, the parenchymatous injection of coloring matter, etc., while the object was not yet cut in sections. For this reason they affected the various hardening fluids that at the same time imparted a tint to the whole material or to certain parts therof. While there are occasions in which this method is yet useful, it has, as a general thing, been replaced by staining the cut sections. Sometimes both processes may be resorted to with advantage in one and the same substance, the latter process being supplementary to the earlier, or mass staining.

§XLV. Since a stain begins to operate the moment that it is brought into contact with a substance, and the operation of tinting progresses as the agent penetrates the tissues; and since the depth of the color depends upon the amount of the agent that is absorbed, which in turn depends upon the length of time that the staining material remains in contact with the tissues, we may propound the following as AN AXIOM.

Whatever agent be employed, the best results are attained when it is used in a diluted form and allowed to operate slowly.

There are, of course, exceptions to this rule, very good results being sometimes obtained from the use of concentrated stains acting with great rapidity. Such is the case for instance, in staining exceedingly thin sections, or delicate and tenuous membranes.

There are other reasons beside the uniformity in the distribution of color attained by the slower process, why an evenly tinted section, less vividly stained, is preferable to one more deeply colored, chief of which is that in the examination of such a preparation with the higher powers of the microscope, less light is required for its penetration; the eye is not so readily fatigued, nor is it attracted to the more vivid points to the neglect of equally important, though less highly colored, portions of the mount.

§XLVI. THE AMOUNT OF DILUTION which the various stains that we have described will stand, and the length of time required by them for the proper tinting of material, are largely dependent upon the nature of the latter. No general rule can therefore be given, but the following suggestions will be found useful:

Beale's carmine and borax carmine may be diluted with from 8 to 10 times their bulk of distilled water, and a writer in the Royal Journal of Microscopy recommends for certain purposes the addition of 20 volumes of water.

Indigo-carmine will stand a very large dilution. The best plan is to take about enough distilled water to thoroughly immerse the object and to add to it indigo-carmine sufficient to produce the shade known as sky blue.

Picro-carmine, in the strength already given (10 grains to the ounce) is quite weak enough for all ordinary purposes.

The anilin colors, chloride of gold, nitrate of silver, etc., may be used in the strength indicated in the solutions, or they may be reduced by from 2 to 5 volumes of water or alcohol, as the menstruum may be.

These dilutions should be made in the watch-glasses before the sections are placed in them, for obvious reasons.

§XLVII. SIMPLE STAINING.—If the sections were floated when cut, into water or glycerin, or even in dilute alcohol, they may be moved thence into any of the aqueous stains without any intermediary; but, if an alcoholic stain is to be used they should first be dehydrated by immersion in absolute alcohol. A neglect of this precaution will sometimes result in the precipitation of the stain and the ruin of the sections. This is especially true with regard to some of the anilin blues.

1 *Carmine Staining.*—Dr. Beale gives the following directions for the use of his carmine: "After the section has been stained it is to be washed in a solution consisting of strong glycerin 2 parts and water 1 part. After a lapse of 5 or 6 hours or more it should be transferred to the following fluid, viz: strong glycerin 1 oz., strong acetic acid 5 drops. After having remained in this fluid 3 or 4 days it will be found that the portions of even soft and pulpy textures will have regained the volume which they

occupied when fresh." (See Microscope in Medicine, p. 66).

Excellent results may be obtained with any of the carmines without so long a delay by resort to either of the 'fixing solutions, given in §XXXVIII. After the requisite depth of color has been acquired the sections should be removed from the stain directly into the fixer and immersed for a few moments, care being taken not to allow them to remain long enough for the acids to bleach them superficially. From the fixer, remove to distilled water, rinse and return to glycerin or alcohol as the case may be.

Dr. Stowell, of Ann Arbor, in his admirable Manual of Histology. and again in the *Microscope* for July, 1887. p. 181. says "all carmine stained preparations are improved by removing the excess of staining with water and placing them in a one per cent solution of acetic acid. The latter takes out superfluous staining, fixes the staining in the nuclei and brightens it everywhere."

2. *Logwood Stained Sections* do not require the acid fixing agents. If they have been overstained, however, any of the "fixers" will remove the superfluous color. After the sections have acquired the requisite depth of tint (which can only be ascertained by removing one occasionally and examining it), they should be rinsed in distilled water and placed directly into alcohol.

3. The *Anilin Stains* may be fixed so that they will retain their brilliancy for a long time by immersing them in a fluid consisting of equal volumes of a saturated aqueous solution of tannic acid and distilled water. This 'fixer' will decompose if made up in bulk, unless some anti-ferment, such as carbolic acid, be added to it. After immersion in this fluid the sections should be rinsed and placed in a one and a half per cent. solution of tartrate of antimony and potassium in distilled water.

§XLVIII. MULTIPLE STAINING.—By referring to the table given above (§XLII) it will be seen that very few of the stains used in microscopy act with uniformity on all forms of matter, and that the great body of them are in one way or another classed as *selective*. This selective property is either absolute or modified, and may consist—(*a*) in choosing certain tissues or elements upon which to act, while leaving others unaffected: (*b*) in conferring a certain tint to one element or set of elements and a different one to others; (*c*) or, finally, in acting with greater energy and rapidity upon an element or part thereof than it does upon the balance of the substance to be stained. Upon this selective property in its various modifications, depend the processes known as double or multiple staining, and which have been of such incalculable service in all departments of scientific microscopy, but more especially in biology and histology. It is to works on these branches and to future chapters on special methods that I must now refer those who are seeking information upon the selective preferences of individual stains, since here we can only indicate general methods and results.

1. *Picro-carmine.*—As before stated, this valuable stain possesses the property of coloring certain tissues red and others orange, so that in itself it is a veritable double stain. By taking advantage of this fact it is made the basis of several multiple staining combinations. Thus, by immersing a section from the base of the tongue (say of a dog or cat) first in the picro-carmine, and successively thereafter in magenta (rose anilin) and iodine green, the result will be that the muscles are stained by the first, the connective tissue and protoplasmic cells by the second, while the third will color the nuclei in the superficial epithelium, the serous glands and nonstriated muscular fibers in the vessels. Nor is this all—the mucous glands will show a purple from the combined action of the red and green. Picro-carmine acts well with logwood, iodine green and anilin-blue—the method being the same in each *i. e*: stain with picro-carmine, rinse and apply the second color.

2. *Eosin* is another most excellent base or ground stain in multiple staining. When used for this purpose the standard solution should be diluted with 8 or 10 volumes of water and the object left in it but a few seconds. After withdrawal it should be immersed in water containing 20 minims of acetic acid to the ounce, rinsed, and the second color applied. Any of the blues combine well with it.

3. *Logwood.* Hæmatoxylon combines well with picro-carmine, the carmines, anilin blue and the metallic stains. An object first stained with it may be successively stained with any of the others.

4. *The carmines.* Objects that have been stained with any of this series may subsequently be stained with indigo-carmine by removing them from the fixing bath into dilute alcohol, letting them remain there four or five hours and thence transferring

them to the dilute sulph-indigotate of sodium solution. The length of time necessary to obtain the best effects varies from 2 to 10 hours, according to the nature of the section.

§XLIX. There are many other combinations beside those suggested and, as the subject is a fascinating one, the student will no doubt find them out for himself. I have gone into the matter at present as fully as is necessary, or as space will admit, and will conclude by quoting two axioms from Dr. Carpenter:

1. The effects produced by using one stain after another are generally much better than those obtained by simultaneous staining.

2. The selective action of a second stain is not prevented by previous staining; for the dye which made the latter seems to be more weakly held by the parts which take the former, so as to be displaced by it.

Thus, if a section of the stem of a plant be stained throughout with anilin red and then placed (after washing in alcohol) in a solution of Nicholson's blue, ½ grain to the ounce, the blue will in a short time drive out the red. But by carefully watching the progress, it will be seen that the different tissues change color at different times, the softer tissues being first to give up the red and to absorb the blue. Thus, by stopping the process at the right point, the two kind of cells are beautifully differentiated by their coloring matter. (Journal of the Royal Mic. Soc. 1880, p. 694).

BIBLIOGRAPHY. Among the more valuable treatises on stains and staining, I would refer the readers who desire to pursue the matter further to Poulsen's 'Botanical Micro-chemistry,' Frei's 'Microscopical Technology,' Carpenter's 'The Microscope and its Revelations,' Beale's 'Microscope in Medicine' and 'How to Work with the Microscope.' A most exhaustive treatise on the subject appeared in 1884 and 1885 in the *Zeitschrift für wissenschaftliche Mikroscopie* from the pen of Dr. Hans Gierke. A good translation of these articles appeared in monthly parts, in the *American Monthly Microscopical Journal*, commencing in April 1885, p. 65 (published by Dr. Romyn Hitchcock, Washington. D. C.).

IX.

We have now arrived at a point in our progress toward a completed slide where we must leave, for awhile, the material to be mounted, and turn our attention to the preparation of the glass slip which is to receive the same and preserve it for future examination and reference.

§ LI. Every perfect mount consists, or should consist, of five parts, viz: The slip, or glass plate; the cell wall, or wall of cement surrounding the object and retaining the mounting material; the object; the mounting medium; and the cover-glass. The proper union of these elements constitutes the *slide*. The materials requisite for a slide, beside the object to be mounted, therefore are: glass slips, cover glasses, cements, and mounting media. We will consider them in detail:

§ LII. GLASS SLIPS.—Formerly there was no uniformity in the size or shape of these, every microscopist having them cut to suit the stage of his instrument, the size of his cabinet, or as fancy dictated. Every manufacturer of microscopical accessories and materials followed his own inclination in the matter. The Germans, generally, affected a nearly square rectangle, some two inches long by one and three-quarters broad. The French sent three sizes into the market, the smallest 0.5" wide by 2.25" long, and the largest 1.10" wide by 3.10" long. Of late years, however, since the study of microscopy has become so widely extended, and especially since the establishment of postal microscopical clubs and exchanges, the profession has settled upon an uniform slip, one inch wide by three inches long. This should be of the best and clearest glass, entirely free from bubbles and flaws. The edges should be ground, not merely for the sake of appearance, but because when left unground they scratch and disfigure the stage of the microscope and sometimes cut the fingers of those who handle them. Such slips were formerly quite costly, but the increased demand for them has induced manufacturers to devise methods for their rapid production, and the best ground and polished slips of Chance's glass can now be bought for the price charged for plain (unground) ones five or six years ago.

§ LIII. These slips, as they usually come from the dealer, are greasy and unfit for immediate use. The most expeditious method of cleaning them, where a fire is convenient, is as follows: Put them in the bottom of a clean saucepan and pour over them some concentrated lye or liquor potassæ, cover with cold water and set on the fire to boil After they have boiled for a short time, remove from the fire, let cool, pour off the dirty water, rinse with clear water and wipe dry with a clean cloth. After the slips are wiped they should be wrapped in packages of one dozen, in clean, soft paper and kept in a tight box, where dust cannot get to them. This latter precaution is necessary in cities where the air is constantly full of minute particles of silicious sand and dust, which soon dull the polish of the surface of the slips.

Another method of cleaning is to immerse the slides for twenty-four to forty-eight hours in a solution consisting of four ounces of bichromate of potassium dissolved in a pint of water, to which has been added four ounces of commercial sulphuric acid. This solution (which is the ordinary battery fluid) is also most excellent for cleaning cover-glasses. Many workers immerse their slips and cover-glasses in this fluid and leave them there until they are wanted for use.

§ LIV. Cover-Glasses—These are circular or square pieces of very thin, clear glass purchasable of dealers in microscopical supplies, in four grades and six sizes, running from $\frac{1}{2}$ up to 1 inch in diameter. The grades are determined by the thickness of the glass, No. 0, the thinnest, being from $\frac{1}{250}''$ to $\frac{1}{200}''$ in thickness; No. 1 from $\frac{1}{200}''$ to $\frac{1}{150}''$; No. 2, from $\frac{1}{150}''$ to $\frac{1}{100}''$, and No. 3, from $\frac{1}{100}''$ to $\frac{1}{50}''$.

Square cover-glasses, while convenient for temporary balsam mounts, are neither so sightly nor convenient for permanent ones, and though formerly much affected are now but rarely used, even in Germany, whose microscopists have until very recently refused to avail themselves of those time and labor saving accessories, the microtome and turntable.

The cover-glass may also be bought in plates and cut by the microscopist to suit himself. The labor of cutting is very much lightened and simplified by an ingenious little device for the purpose sold by all dealers in accessories. The price of the cut covers is, however, so reasonable (from $1.50 to $3.00 per ounce, according to thinness) that it will scarcely pay anyone to cut them himself, unless he uses a very large number, or wishes odd sizes.

Cover-glasses should be cleaned, as suggested above, by immersing them for two or three days in the chromate solution, after which they should be rinsed and then deposited in absolute alcohol, where they should remain until required for use.

§ LV. THE CELL.—In mounting very thin sections or very minute objects, when Canada balsam, damar or a similar preservative medium is used, a cell wall is not an absolute necessity; nor is it essential even, that the edges of the cover-glass should be sealed and finished with any cementing material, since the mounting medium itself dries hard and impervious at all points exposed to the atmosphere. The appearance, however, and the absolute durability of the mount are always improved by such a finish. It is therefore best to use it in all cases where the mount is intended to be permanent.

When the object to be mounted is of any appreciable thickness, or when it is excessively fragile and liable to be crushed by the pressure exerted by the natural contraction of the mounting medium in drying, the case is different, and whatsoever medium be employed, a cell wall, for the support of the edges of the cover-glass, becomes an absolute necessity. The reasons for this must be patent to anyone who has ever given the subject a moment's thought. When a supporting wall is not used the object itself must sustain the cover-glass until the mounting medium dries sufficiently hard to perform this function. As the medium shrinks in drying—the amount of shrinkage depending upon the amount of volatile matter contained in it—undue pressure is gradually brought to bear upon both the object and the cover-glass. The result of this pressure is, frequently, distortion of the histological elements, or even crushing of the object (as in the case of delicate frustules of the larger diatoms) and a curving of the cover-glass, which renders it peculiarly liable to fracture by the slightest external pressure. So marked is this distortion of the cover-glass, in many instances, that it absolutely assumes a lenticular shape. In the older text books on microscopical technique various devices were suggested for the avoidance of such contretemps—though, strange to say, the true reason for the distortion seems to have escaped all writers upon the subject. Chief of these ancient devices formerly in vogue, were cells made of different thicknesses of paper, cut or punched out by hand. The introduction of round cover-glasses and the turntable now enable the microscopist to make perfectly uniform cell walls of any desired depth; though as we shall see later on, cells made of wax, india rubber, celuloid and of metal, still play an important part in mounting objects of any considerable thickness, especially for direct examination.

§ LVI. CEMENTS—A good cement should have the following qualities, the importance of which is in the order named:

1. It should not be soluble in the mounting medium, even in the slightest degree.

2. It should attach itself firmly to glass and should dry hard and quickly.

3. It should be tough and not become brittle with age.

4. It should flow freely from the pencil, and not have a tendency to become ' ridgy ' or to run.

5. It should shrink but slightly, or not at all, in drying.

To these we may add that it should keep well in a liquid form, and not alter or deteriorate with age.

From these points it is plain that the ideal cement, one which will fulfill all the requirements enumerated, and under all circumstances, is a desideratum much longed for by microscopists, but one which, I do not hesitate to say, has not yet been discovered. Cements, for instance, which answer perfectly well for glycerin or aqueous mounting media, will dissolve and rapidly succumb when brought in contact with balsam or prepared damar. On the contrary, those which are unaffected by balsam and its solvents (benzol, turpentine, chloroform, etc.) will soften when used with glycerin or watery fluids. No one cement, therefore, with which we are at present acquainted, will answer under all circumstances. The following are the best cements in use at present:

(a). ASPHALT CEMENT—This is one of the best and earliest of the cements employed in cell making, and up to very recently it was used, either alone or in combination with caoutchouc, drop black, or other ingredients, by the best foreign mounters, almost to the exclusion of all other cements. It is known in commerce as *Brunswick black* and, as purchased from dealers in painters supplies, consists of the best foreign asphalt dissolved, with the aid of heat, in linseed oil. Litharge and gum amber are added during the boiling, which is continued until all fluids are driven off and a nearly solid mass is obtained. This mass, dissolved in oil of turpentine, constitutes the asphalt varnish of commerce. The varnish may be used alone for cell making, but a far better cement results when the solid asphalt, prepared as above, is dissolved in twice its weight of benzol, in every ounce of which four grains of caoutchouc have been previously dissolved. Black Japan varnish is sometimes used in place of Brunswick black. One of the best articles of this sort that has ever come under my notice, and one which presents when finished an appearance identical with the cell material used by the celebrated French mounter, Bourcogne, of Paris, is a black enamel leather dressing used by carriage finishers. It comes in square tin flasks, holding one pound, and may be obtained from any dealer in coach makers' supplies.

(*b*). GOLD SIZE. This is another most excellent cell building and finishing cement. The kind used is that technically known among gilders as oil size, and used for gilding outside work which has to stand the action of the weather. It may be obtained ready-made from any dealer in artists' or painters' supplies. There is much difference, however, in the quality of the material as sold by dealers in microscopical accessories—a great deal of it being nothing but boiled linseed oil. The following is the process for preparing the best: Grind together twenty-five parts of strained or filtered linseed oil, one part calcined red ochre and one part red lead, and boil for three hours. Let stand until the solid matter sinks to the bottom and decant the clear fluid. Mix the latter with an equal bulk of white lead, boil for an hour and again let stand and decant. The clear liquid is a size that attaches to glass with wonderful tenacity and dries as hard as flint. It will stand much rough usage and is altogether one of the very best of cements.

(*c*). SHELLAC CEMENT. Break white shellac into small pieces and soak in sulphuric ether over night. In the morning pour off the surplus ether, drain completely and dissolve in absolute alcohol. Bleached shellac dissolves very slowly, even with the previous softening in ether, and it requires a long maceration to obtain a good solution. A considerable amount will remain undissolved after weeks of contact with absolute alcohol. To obtain a limpid product, add to the solution about one half its volume of benzol, agitate and let stand. In a few hours the liquids will separate, the benzol retaining the detritus and leaving a layer of perfectly clear alcoholic solution of shellac. This makes a very beautiful and durable cell. It may be colored to the taste by the addition of any of the anilin dyes that are soluble in alcohol.

(*d*). OXIDE OF ZINC CEMENT. To prepare this cement well requires some labor and care, which, however, are in my opinion amply repaid by the result, since for excellence and general utility this cement is inapproachable by any other yet invented. The materials are gum damar, benzol, oxide of zinc, and a small amount of some drying oil, preferably nut or poppy oil. The damar is dissolved in benzol to the consistency of a thin syrup, and the solution carefully filtered through absorbent cotton. The zinc oxide should be white and perfectly free from moisture, to ensure which I am in the habit of drying it thoroughly by heating in a muffle. The zinc changes color, becoming yellow under heat, but soon regains its whiteness. In preparing the cement only a small portion of the zinc should be put into the mortar at once. This should first be wet with a few drops of benzol, the solution of damar then added, and the whole carefully rubbed until a smooth and intimate incorporation of the ingredients is effected. The mixture is then poured into a stock bottle and the operation repeated until a sufficient amount of the cement has been made. In despite of the most careful and continued use of the pestle or muller, some portion of the zinc will remain uncrushed, and to get rid of these particles the fluid must be filtered through absorbent cotton or allowed to stand awhile and then decanted. The latter is far the best and most economical process. The larger particles sink more rapidly than the finer ones, and in the course of two or three decantations an exquisitely smooth and homogeneous fluid is obtained.

To get the best results in practice, the proportions of zinc and damar solution, in the finished cement, should be about equal. To secure this result let the decanted fluid stand until all of the zinc has separated and fallen to the bottom of the vessel. If there is not enough of the damar solution, more can be added from the stock bottle. If too much be present, the surplus may be drawn off with a pipette. The cement is completed by the addition of from fifteen to twenty minims of drying oil to each ounce of cement. Without the latter ingredient the product becomes brittle on drying.

As thus made, and used with the precautions hereafter indicated, there is in my opinion no cement which can compare with it, either in the smoothness with which it flows from the pencil, rapidity of drying, tenacity and evenness of attachment, or the hardness and toughness of the product. In addition to these practical recommendations, there is nothing in the way of cements that is more beautiful in finish. In saying this much about the cement I am quite aware that it has been decried and abused by certain parties, belonging to that class of workmen who invariably lay their own lack of skill upon the tools they use. I have used it constantly for fifteen years and whenever it has failed to make a good and durable mount the failure could be traced directly to my own negligence or carelessness.

(*e*). MARINE GLUE.—Dissolve one part of caoutchouc in 32 parts of coal tar naphtha, and when solution is complete (in the course of 8 or 10 days) add $2\frac{1}{2}$ parts of dark shellac in fine powder and let the vessel, closely stoppered, stand for another week or ten days. At the expiration of this time put it into a sand-bath and heat carefully until the undissolved portions of shellac and caoutchouc are liquefied and a homogeneous fluid is obtained. This should be poured upon a cold plate or slab and allowed to solidify, after which it may be broken into pieces and preserved for use. It requires a temperature of $240°$ F. to liquefy this cement. It is the most tenacious of all substances yet devised for this purpose.

A liquid cement, resembling the foregoing in general characteristics, may be made by dissolving 1 part of india rubber in 12 parts of benzol and adding to the solution 20 parts of shellac. Heat in a water bath until solution is complete. This cement may be applied with a pencil—which that made by the former process cannot be, unless redissolved in benzol or naphtha.

(*f*). BELL'S CEMENT is the proprietary name of the shellac solution given above (*c*).

(*g*). DIAMOND CEMENT.—Dissolve one ounce of best Russian isinglass in $5\frac{1}{2}$ ounces of distilled vinegar. In a separate vessel dissolve a half ounce each of gum ammoniac and gum mastic in two ounces of alcohol and mix the two solutions. This cement is very tenacious and is used by Oriental jewelers in setting precious stones.

SEILER'S CEMENT is a modification of the above. It is made by dissolving two drams of isinglass in one ounce of acetic acid in which ten grains of gum ammoniac have been previously dissolved.

A similar cement may be made by dissolving one ounce of isinglass in four ounces of hot skimmed milk.

(*h*). CASEIN CEMENT is made by dissolving curd or pot cheese made from skimmed milk, in a hot saturated solution of borax. It makes a hard and brilliant cement, useful in certain cases.

(*i*). ARABICIN CEMENT.—To a thick solution of gum arabic add sufficient alcohol to throw down the arabicin in the form of a white flaky deposit. Throw on a filter and wash with alcohol, dry and redissolve in distilled water to the consistency of an officinal syrup. To each ounce of the solution add twenty-four grains of sulphate of aluminium and dissolve. With this as an excipient, rub in sufficient finely pulverized talc to make a fluid of about the same consistency as good zinc cement. This cement may be made directly from a solution of gum arabic, but it does not keep so well as when the arabicin alone is used. It is useful for building cells for balsam mounts, answering the same purpose in these cases as zinc cement does when glycerin or aqueous mounting media are used.

As in the case of stains, formulæ for cements might be multiplied almost *ad infinitum* were any good purpose subserved by it. With those given above the student can meet all the exigencies of microscopical technology, and any more would be superfluous.

THE CELL WALL

X

§. LVII. As hitherto intimated, the choice of a cement for building the supporting wall of, and closing the completed cell, depends very much upon the nature of the mounting medium to be used. To be permanent it is manifest that the container must be absolutely insoluble in the fluid which it is to hold and with which it remains in contact. Hence, for technical purposes, we may divide mounting media into two great classes, viz: the aqueous and the gummy or resinous. In the first we place glycerin, glycerin jelly, camphor water, camphorated gelatin, carbolized gelatin, serum, etc., soluble in water. The second comprises Canada balsam, damar, the oils, etc., insoluble in water, but soluble in benzol, chloroform, oil of turpentine, ether, alcohol, etc.,— fluids which are used in preparing most of the cements used in microscopy. At first sight it would seem that no great difficulty should arise on this score and that the trouble would be obviated by using aqueous cements for gummy or resinous media, and *vice versa*. But in practice we find that the aqueous cements are attacked by the moisture of the atmosphere, by the water used in cleansing slides (which will occasionally get soiled) and finally, by the fluids used with immersion lenses. This difficulty is overcome by covering the aqueous cements, after they have set and dried thoroughly, by a resinous cement, impervious to and unaffected by the atmosphere. An analogous operation enables us to protect a resinous cell wall from being attacked by similar mounting media. In such cases all that is necessary is to cover the wall with an aqueous cement which, when dry, acts as a complete protection to it. This enables us to use one and the same cement for the foundation of all cells, whether as containers for balsam or glycerin, damar or gelatin, leaving subsequent manipulations to be determined by the mounting medium to be employed. The importance of this apparently trivial point is that it enables the microscopist, to keep on hand at all times a stock of thoroughly seasoned cells ready to be filled when wanted—a point which will be recurred to later on.

§. LVIII. THE TURN TABLE.—Of all the instruments which of late years have been introduced as aids in microscopical technology, not one has conduced so much to economy of time, permanence of mount and elegance of finish as has the turn-table. In its simplest form this is a disc of heavy wood or metal, three and a half or four inches in diameter, mounted so as to rotate easily upon a perpendicular pivot or axis, and provided with clips to hold the slide in place. A little table, somewhat larger than the disc, acts as a hand-rest in operating. Fig. 7, represents such a turn-table.

FIG. 7. TURN-TABLE.

Upon this simple instrument there have been constructed quite a number of improved turn-tables—the improvements consisting in devices for the automatic centering of the slip, improved methods of holding the same, devices for decentering, etc. Nearly every manufacturer of microscopical accessories has some such improved turn-table, and many of them are very meritorious. Among the best of the self-centering devices I may refer to those of Griffith, Bausch & Lomb, Bullock, Walmsley, and Queen.

FIG. 8. GRIFFITH'S TURN-TABLE.

Fig. 8 represents the self-centering and improved turn-table of Mr. E. H. Griffith, of Fairport, N. Y., manufactured for him by the Bausch & Lomb Optical Company, Rochester N. Y.

§. LIX. METHODS OF USING.—Having placed the slip, thoroughly cleaned and dried, upon the turn-table, and centered it by the eye (if the instrument is not automatic), the disc is given a rapid rotary motion by laying the forefinger of the left hand upon the milled edge of the disc itself or the button underneath it, and drawing it quickly toward the person. In the meantime, the right hand, holding a camel's hair pencil which has been dipped in cement, is rested upon the "table" and the point of the pencil is brought into light contact with the surface of the rotating slip. The brush should be held exactly as a pen in writing, and care must be taken that the contact of the point shall be tangential to the diameter of the circle that it is desired to make, and at right angles to an imaginary line drawn from the body of the operator through the center of the rotating disc. This is in order that the friction of contact shall draw the hairs composing the pencil in a straight line, away from the hand and parallel with the handle of the brush. If held otherwise the point will be twisted and the circle made smaller or larger than is desired. A very little practice will give the 'knack' and will teach the student just how much cement to take up with the pencil.

§. LX. The cell wall, as usually built, should have a width of about one-sixth of an inch, and the outer diameter should project about half that space beyond the edge of the cover-glass, all around. The depth of the cell depends upon the thickness of the object that it is desired to mount in it. For ordinary sections of pathological or histological material, the depth should not be over one five-hundredth part of an inch– which is obtained by putting on from two to three layers of oxide of zinc cement. When more layers are required the first two or three should be allowed to dry quite thoroughly before adding others. When the cells are to be very deep other devices and precautions will be necessary, and will be described under a special heading.

§. LXI. Many microscopists are in the habit of making their cells as they need them, allowing the rings to dry only so much that the cover-glass will not stick fast when it is applied. Some do this from ignorance and thoughtlessness, while others, who have never experimented upon the relative durability of a cell made all-at-once and one constructed upon a thoroughly dried and seasoned wall, claim actual advantages for the former procedure. They say when the cover-glass is applied to the walls while they are yet plastic, a more accurate coäptation of surface is obtained, and a more homogeneous mass is made with the cement that is afterwards applied in closing the cell. These advantages, if they be such, are more than counterbalanced, in all except dry mounts, by a radical defect inherent in all such hastily prepared mounts, whether made of zinc white cement, Brunswick black, or any other material with which we are acquainted, viz: *due allowance cannot be made for the shrinkage of the cell in drying.* This is the secret of most of the failures and disappointments which produce the bitter complaints that we find in the technical journals, from correspondents denouncing this or that cement or mounting material.

All the cements described in the foregoing chapter, with the exception of gold size, consist of a solid material or materials suspended or held in solution in some more or less volatile medium the evaporation of which again leaves a solid mass. The exception (gold size) hardens partly, though very slightly, by evaporation, its solidification depending principally upon a chemical change wrought by the oxygen of the atmosphere. But even this change is accompanied by a diminution of volume; and as to the cements composed partly of volatile material, it is plain that there must be a very large decrease of volume in the process of solidification. In asphalt, zinc-white and shellac cements and marine glue among the resinous cements, and arabicin and gelatin cements among the aqueous, this shrinkage amounts to a diminution of volume of at least 30 per cent. When a cell is properly finished it must be entirely filled with the mounting medium. If it is not so filled we are bound to have air bubbles,—which are not only unsightly but which ultimately destroy the mount. It is plain that if we entirely fill a cell with any mounting medium and this cell afterwards loses one-fourth or one-third of its volume, something must give way. The fluid (air excepted) is practically incompressible, yet great pressure is brought upon it. It has no space within the cell into which it can retreat, and consequently it must force its way out of it. The pressure is slow, and gradual, but continuous, and finally the cell gives way at its weakest point; the medium exudes or 'creeps out' and is discovered. It is washed off and more cement applied. In a few months the process is repeated—the fluid gradually infiltrating and disintegrating the cement, until finally the slide is a total wreck. The builder of it, meanwhile, not suspecting that he has undertaken the old, old problem of opposing an unyielding body to an irresistable force, damns the cements or the mounting medium, or both, and forthwith indites a communication to some technical journal, the editor of which 'sympathizes' with him, and the twain unite in solemnly warning the profession not to use the offending materials "unless they are prepared to have a certain percentage of their mounts destroyed." This is another example of the workman blaming his tools for his own lack of skill. If a 'certain percentage,' only, of mounts made with a cement or mounting medium are ruined by any means whatever, it is proof positive that a certain percentage remains which is not thus affected. And since the operations of nature are carried on by fixed laws, and not by caprice of inanimate matter, it is self-evident that those which were destroyed lacked the skill and care that was expended on the balance. All of which induces us to formulate and propound the following:

AXIOM.—Never finish a cell the walls of which are not thoroughly dry and seasoned.

§. LXII. ANOTHER PRECAUTION which I would impress upon the beginner is the absolute necessity of having his glass slips clean and, above all, free from moisture, especially when balsamic, oleaginous or resinous cements or mounting media are to be employed. It is really wonderful how small an amount of moisture will destroy the tenacity of the one and the transparency of the other. And the same precaution against moisture should be observed in making the solutions of damar and other resins and gums, in preparing the cements. The few drops of water adherent to the sides of a bottle will render milky and opaque a pint of the solution of damar in benzol.

§. LXIII. CELLS FOR DRY MOUNTS, or mounts in which atmospheric air is the medium, and which are almost always intended for direct examination, are usually required to be much deeper than those for examination with transmitted light. Instead of resorting to the slow process of building them with cement, layer by layer, they may be quickly prepared by cementing to the slide rings made of glass, metal, vulcanite, ivory, mother of pearl, in fact of almost any solid material of a proper size and shape. Brass curtain rings, a pearl button with the center scooped out, the eyelet of a copper rivet, all make excellent cell building material. They may be attached to the glass by any of the cements given above, but I prefer marine glue for the purpose. Beautiful cells for this class of mounts may be made of sheet wax, as suggested by the Rev. Dr. J. T. Brownell at the Rochester meeting of the American Society of Microscopists (1884). The wax used for the purpose is that made for flower workers. Several layers may be superimposed, and the whole can then be turned into shape by placing the slip upon the turn-table and applying the edge of a sharp knife or chisel to it as it revolves.

At the Chicago meeting of the Society in 1883, Prof. A. H. Chester exhibited some very handsome cells made by punching annular discs from thick tinfoil and cementing them to the slip with marine glue.

A very beautiful cell is made of shellac in the manner described by Mr. Rebaz before the same society at the Rochester meeting. The details of these three processes are too long to be described here, but may be found in full in the reports of the Proceedings of the A. S. M. for 1883 and 1884.

Finally, a most excellent cell may be made from a paste of litharge and glycerin. The paste is applied to the center of the slide and as it commences to set the slip is placed upon the turn-table and the mass turned down to a proper size and shape. A skillful use of the turn-table, after the fashion of a lathe, enables one to make cells of almost any plastic material.

MOUNTING MEDIA.

XI.

The series of operations preliminary to the final act of mounting have now been completed. They have thus far progressed in separate but parallel lines, the one having to do with the preparation of the object to be mounted, and the other with making ready the glass slip which is to receive it. We have now arrived at a point where the manipulations will be with both, for nothing now remains but to place the object within the cell-wall, arrange it as we wish to have it remain, surround it with the proper preserving medium, adjust the cover glass and cement it to its proper place and our task is completed—we have a finished slide. But in this series of manipulations, simple as they seem, are embraced some of the most difficult and important points in microscopical technique. Our slide being ringed and clean, and our object ready, the first question that demands our attention is the choice of a mounting medium.

§LXIV. MOUNTING MEDIA.—By this technical term we designate the fluids or materials with which the objects are surrounded and permeated and the cell filled. This medium has a two-fold function, viz: preservative and optical. The first is self explanatory and the second has already been explained in a previous chapter (§VI), to which we must now make reference.

§LXV. Like cements, mounting media may be broadly divided into two classes, viz: the resinous and the aqueous. The first embraces Canada balsam, gum-damar, liquidambar, the drying oils. etc.; the second, glycerin, gelatin, camphor-water, glycerin jelly, etc. The choice of a medium will depend in most instances upon the nature of the object to be mounted and the character of the investigations to be made. These problems belong to special technology and hence need not be discussed here, except in a general way. Leaving out those cases where optical problems must be worked out (as, for instance, where a medium of a certain definite refractive index is required), or those cases where certain peculiar preservative properties are demanded, the choice of a medium is practically limited to two or three substances, viz: Canada balsam, glycerin or gum-damar, and the student will rarely go wrong when he chooses one of these media. They are commonly,—I might say almost universally, used among microscopists for histological and pathological investigations.

The methods of preparing and using the balsam and damar are essentially the same, while the manipulations of glycerin may stand as a type of mounting with aqueous media.

(*a*). CANADA BALSAM.—This is the oldest and best known of all the mounting media, having been used for this purpose from an early period in the history of the science. At first it was used in its crude state—that of a semi-fluid terebinthinous resin, the exudation of a tree belonging to the family of firs (*Abies Balsamea*) and the mounts thus made, owing to the large proportion of oil of turpentine and other fluids contained in the resin, took a very long time to set and harden. For many months after preparation the slightest touch would cause a displacement of the mount. This defect is remedied by driving off the turpentine and other volatile matters by exposing the crude balsam to prolonged heating in the water-bath, a process which converts it into a solid, brittle, resinous mass, closely resembling ordinary resin or colophony. This mass, dissolved in chloroform or benzol to a proper consistency, constitutes the "balsam" now almost universally used by microscopists. It may be purchased ready prepared from the dealers in microscopical supplies.

(*b*). GUM-DAMAR.—This gum in its crude commercial state consists of hard, transparent, brittle nodules varying in size from that of a pea, up to a hickory nut. It is also an exudation from a tree belonging to the fir family, and known as the damar or damara pine. It is prepared for use by dissolving in chloroform or benzol. As the crude gum always contains more or less dirt, fragments of bark, etc., the solution must be clarified for use. The larger particles of foreign matter may be removed by running the solution through absorbent cotton. To get rid of the finer dust is a more difficult matter. A number of plans for clarifying it have been suggested but about the best and most convenient (though not the cheapest) is to make the solution so thin that it will pass through filter-paper. This makes it perfectly limpid, and the surplus benzol or chloroform may be afterwards driven off by heat. Damar by itself soon becomes very brittle and somewhat opaque: these faults are remedied by adding to the solution prior to filtration, a few drops of nut, poppy, or linseed oil.

A perfectly limpid and colorless solution of damar, of high refractive index and great beauty may be made as follows: To the clarified solution in benzol add alcohol of 90° until a precipitate is no longer formed. Remove the precipitated gum and wash with distilled water and afterward with alcohol; let dry thoroughly, and redissolve in pure benzol. This resin, when dry is exceedingly brittle, falling into an impalpable white powder upon the slightest pressure. The addition of 20 drops of poppy or nut oil, while imparting a faint yellowish tinge, corrects the brittleness. Mounts made with this medium never become discolored or opaque—or at least some that I have kept four years and have constantly exposed to the light, still remain absolutely colorless and transparent, while balsam mounts made at the same time and submitted to the same influences, are quite yellow. If the student has facilities for so doing, it will repay him to recover the excess of benzol, alcohol, chloroform, etc., used in making these preparations, by the usual processes of distillation at low temperature.

(*c*). LIQUIDAMBAR, styrax, or 'sweet gum' is another resin of high refractive power and though somewhat high colored, is valuable as a mounting medium. It is prepared and used like Canada balsam. It has been found especially valuable as a medium for mounting diatoms.

(*d*). MASTIC.—This is a resinous exudation from a shrubby plant of the pine family, the well-known *pistacia lentiscus*, of the Grecian Archipelago. It is freely soluble in chloroform, ether, benzol and turpentine. Alcohol dissolves about four-fifths of it, leaving a friable, brittle mass called masticin. Although apparently well-fitted for a mounting medium mastic is but little used for this purpose. Some mounts made with it four years ago are now in excellent condition. Its refractive index is about the same as that of Canada balsam.

(*e*). COLOPHONY, or common pine resin, is another substance that is apparently well fitted for use as a mounting medium, and which is scarcely ever used—probably because it is so plentiful and common. Thiersch and Frei are both loud in their praises of this material, the former preferring it to Canada balsam. To prepare for use, pick out a lump of the clearest virgin resin and dissolve it in sulphuric ether, filter and heat the filtrate in a waterbath until the ether and the originally contained oil of turpentine are driven off, and the residual resin when cold breaks with a clean, conchoidal fracture. This should be redissolved in absolute alcohol, and the solution used in the same manner as Canada balsam.

This completes the resinous media commonly in use. The method of applying them will be explained further on.

§LXVI. AQUEOUS MEDIA.—Under this term I include all media that are soluble to any extent in water, or which readily mingle with the same and give clear resultant fluids. A very large number of such media have been suggested from time to time, but we need now consider only those which have been found to answer all general purposes in histological and pathological work. My own experience with the great mass of the suggested materials has convinced me that except in certain rare cases all except those that I mention below can be dispensed with without any serious inconvenience.

(a). GLYCERIN.—This fluid comes at the very head of the list of those media which may be used under almost all circumstances When pure it is a colorless, odorless fluid, of a syrupy consistence. oily to the touch. and very hygroscopic—absorbing water with great avidity from everything with which it comes in contact and which contains that fluid. It is neutral in its reactions, soluble in water and alcohol in all proportions, but insoluble in ether. chloroform, turpentine, or benzol, or the fixed oils. Its specific gravity is 1.225, and its refractive index 1.475. Its high refractive power, its affinity for water, its preservative properties. and its stability combine to make it the most valuable of all the media for mounting animal and vegetable tissues containing water. As remarked by Frei, "what balsam is to dry tissues, glycerin is to moist ones." Possessing antiseptic properties of no mean order. it may be used alone or its preservative properties may be reinforced with carbolic, acetic, or formic acid. Since pure glycerin renders some delicate tissues too transparent for accurate examination, it sometimes becomes necessary to reduce this property by the addition of distilled water or alcohol. No definite rule can be given for this dilution, but since it is advisable that those objects destined to be mounted in it, should remain in glycerin for several days before they are finally transferred to the slide, the student may easily satisfy himself on this score before finally finishing the mount. Where only a slight amount of dilution is necessary the operator may safely use the pure glycerin, relying on sufficient water entering the cell beneath the cover-glass in the act of washing the unfinished cell.

Dr. Beale recommends that a small amount of acetic acid be always added to the glycerin, especially in mounting objects that have been injected with his blue or carmine fluids. Frei suggests hydrochloric acid, and Ranvier formic acid for this purpose. One part of acid to 100 of glycerin is abundant in any case, and where hydrochloric acid is used the proportion should not be over one part to 500, or even weaker. For objects not injected or stained Professor Bastian strongly recommends 6 parts of crystalized carbolic acid to 100 of glycerin.

(*b*). GLYCERIN JELLY.—Soak a good article of gelatin in cold water until it has taken up as much of the fluid as it will absorb. Throw on a coarse cloth or seive and drain off all superfluous water. Transfer to a water bath and heat until the gelatin melts into a homogeneous fluid. To this add an equal volume of pure glycerin, and continue the heat until the water bath boils. Strain the mixture through a clean piece of white canton flannel and preserve for future use. This jelly will keep indefinitely and when required for use needs only to be fluidified by immersing the container in moderately warm water. It is to be used in mounting the same class of materials as pure glycerin.

(*c*). CAMPHOR WATER is made by placing a lump of camphor is distilled water and letting it remain until the fluid has a strong camphoric odor and taste. Though much praised by some (especially English) workers. I have rarely found any use for this medium. Creosote water, used for the same purposes as camphor water, is made by dissolving three parts of creosote in a fluid consisting of 3 parts alcohol and 94 parts of distilled water.

Methylated spirits, alcohol and glycerin, alcohol and carbolic acid, glycerin and honey, serum, and a thousand other mixtures and simples have been suggested as media possessing real or fancied advantages for general or special mounting purposes. An experience of nearly twenty years has taught me, however, that balsam, damar and glycerin (alone or in combination as above) are practically sufficient for all emergencies in mounting, except, of course, where a medium of exceptionally high refractive index is required in some special investigations—such, for instance, as Prof. Hamilton Smith's high index media for mounting the diatomaceæ. These special media will be given in another part of this work.

MOUNTING IN BALSAMIC MEDIA.

XII.

§. LXVII. The student who has followed this series will remember that when we finished section cutting our sections were left in a watery medium, viz; plain water, alcohol and water or glycerin and water, as the case happened to be. Since the sections are therefore permeated with a medium which will not mingle with any of the balsamic mounting media, our first care must be to get rid of it, and to supplant it with a fluid that is soluble in the balsamic medium of which we intend to make use. This process is called *dhydation*, and since there is no known fluid, that can be used for this purpose, which is at once soluble in water and in the balsams, the operation must be performed in two or more stages, or by the aid of two intermediaries, instead of one, viz:

(*a*) Alcohol, soluble in water and essential oils:

(*b*) An essential oil, soluble in alcohol and in the balsamic medium.

§. LXVIII. The Manipulations are sufficiently simple. The section, or object to be mounted, is removed from the watery medium in which it has hitherto rested, and transferred to a watch-glass or other vessel containing alcohol of 95 per cent. After letting it remain in this bath long enough for the alcohol to permeate the tissues, the fluid is drained off and the vessel filled with absolute alcohol. In order that the strength of the alcohol shall not be perceptibly reduced by the aqueous matter extracted from the material introduced therein, the volume of the fluid should be large in proportion to that of the material. When the sections are large or comparatively bulky, it is better that they pass through three or even four washings with alcohol before being transferred to the essential oil. The length of time requisite to this end depends upon the nature of the material, but need rarely occupy more than five or ten minutes.

§. LXIX. When the object is permeated with alcohol it should be thoroughly drained and then placed in some essential oil, which as remarked above, is equally soluble in alcohol and the balsamic mounting medium. One of the best known oils and that most frequently used for this purpose, is oil of cloves, although there are a number equally as good and some of them free from the objectionable characteristics of clove oil (such as shriveling of tissues, pungency of odor etc.). Among the agents which I have successfully used are oil of cade, oil of turpentine, creasote and carbolic acid. Whatever oil be used, the object must be left in it until thoroughly permeated, a fact of which the student must assure himself by examining his specimen under the microscope before its final removal into the mounting medium.

§. LXX. When at length the alcohol is completely supplanted, the last step may be taken, and the object placed into the resinous medium in which it is to be permanently mounted. For this purpose I keep on hand a solution of balsam or damar in benzol or chloroform, made much thinner than the balsam used for mounting, and into this the specimens are transferred from the essential oil and left to soak until ready to be transferred to the slide.

§. LXXI. IN MOUNTING WITH BALSAM the operator has the choice of mounting with or without a cell wall. If the object be very thin the latter method is preferable, inasmuch as it requires less manipulation; but since the balsamic mounting medium is liable to shrink very considerably in the process of drying and hardening, I would not advise the student to dispense with the cell-wall whenever the object to be mounted is delicate and friable (as, for instance, the delicate frustules of the larger diatoms). In this method of mounting the manipulations are very simple. The object is removed from the thinner balsamic solution (as little of which as possible should be carried with it) and is arranged in the position which it is to occupy upon the slide. A drop of the mounting balsam is next placed upon the centre of a coverglass which has previously been cleaned and slightly warmed. With a quick movement the cover-glass is turned over and applied to the object, being carried to its place with a very gentle pressure. The slide is now placed on its edge in a warm place and allowed to stand until sufficiently hardened for subsequent manipulation. Air bubbles, the source of so much trouble in other kinds of mounting, soon dispose of themselves by gradually making their way to the edges and there breaking.

§. LXXII. MOUNTING IN BALSAM WITH A CELL.—The method of spinning the cell wall has already been described, and the student has been cautioned as to the choice of a cell material. If any of the balsamic cements (like white zinc, asphalt,etc)have been used for the cell it will be necessary to protect it thoroughly by a coating of some aqueous cement (such as gelatin or arabicin) applied with a camels-hair pencil. If the arabicin cement (described in § LVI, e.) be used for the cell wall no further precautions are necessary and the method of procedure is identical with that described in the foregoing paragraph. The object is arranged within the ring, the balsam applied to the cover-glass and the latter deftly turned and shoved home.

For applying balsam to the cover-glass or object, nothing is better than a bit of glass tubing drawn out to a somewhat coarse point, or a little medicine dropper with a vulcanized bulb.

The beginner is prone to get too much balsam on his slide, an error which causes some loss of time, to say nothing of its being wasteful and slovenly. It is better, however, for him to err in this direction than in the opposite, or in getting too little. In the former case the excess may easily be scraped off after it is dry, while in the latter, considerable manipulation is sometimes necessary to remedy the default. The best plan is to place a drop of balsam at the edge of the cover-glass on the side opposite to the deficit. A needle is then inserted under the cover-glass and the latter slightly raised. As it raises the added balsam is drawn under it and a little manœuvering suffices to distribute the fluid evenly over the field.

§. LXXIII. HARD BALSAM was formerly much used for mounting, but of late it has given way, in a very large measure, to balsam dissolved in benzol or chloroform. There are, however, instances in which it may be used advantageously, and when such is the case it is softened by the application of gentle heat, applied by means of a water-bath. The slides and cover-glasses are also warmed. The manipulations otherwise are the same as in mounting with soft balsam.

When, on the c ntrary, the natural balsam, or balsam from which the turpentine has not been thoroughly eliminated, is used the process of drying becomes a very tedious one and may occupy many weeks or even months. It may be hastened by the application of artificial heat, and a number of devices have been contrived for this purpose. The simplest and best is a rack, made of tin or wood, in which the slide may be packed and hung up in proximity to the domestic heating apparatus, but not near enough to the fire to injure the specimens.

§LXXIV. Finishing Balsam Mounts.—When the surplus balsam around the edge of the cover-glass has become hard and brittle, the slides are ready for finishing. The first step is the removal of this surplus, by scraping with a penknife or other suitable instrument. Care must be taken, in doing this, not to get the point of the blade under the edge of the cover-glass, or in any way to disturb the same; for while the exuded resin may be dry and hard, that which is under the cover is probably still soft and fluid. For the same reason the operator must be careful about making pressure on the cover-glass, as in this manner a portion of the soft balsam may be forced out and its place taken, too frequently, by an air bubble which it is almost impossible to get rid of. If such an accident should happen it may sometimes be remedied by proceeding as suggested in §LXXII; but before fresh balsam is added in such a case, the slide in that neighborhood must be made clean as possible, since the balsam in entering will carry along with it any particles of dust or dirt with which it may come in contact. After refilling, the slide must be again laid away to recommence the process of hardening. After scraping away as much of the dry balsam, etc., as possible, place the slide on the turn-table and spin a ring of arabicin or gelatin cement around the edge of the cover-glass. Let dry, and as soon as this occurs, clean the slides thoroughly with a linen rag moistened with benzol or turpentine. The ring of arabicin prevents the cleansing fluid from invading the cell. After the entire slide and the cover-glass are cleaned, the ring of arabicin cement may be rinsed off with clear water and the slide labelled and put away, or it may be finished as hereafter described in the manipulations of aqueous mounts.

MOUNTING IN AQUEOUS MEDIA.

XIII.

§ LXXV. Before going into the details of mounting with glycerin, which may be taken as the type of the aqueous media, it will facilitate matters to describe here certain simple forms of apparatus which the student can make for himself and which will greatly assist him in subsequent operations.

(*a*) The Glycerin Bottle. This should hold about 2 ounces and should be provided with mouth and delivery tubes fashioned after those of a wash bottle. The mouth or blow tube should be curved in a complete loop before entering the cork, in order to prevent the entrance of dust, etc. A vulcanized rubber bulb, such as is used on atomizers, attached to the mouth tube is a great convenience, but not an absolute necessity. When thus arranged the apparatus forms a little pump which delivers a drop of glycerin at any desired point, and the fluid is in the mean-time kept free from dirt and extraneous matters.

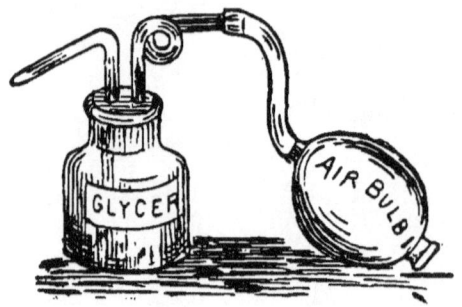

Fig. 9. Glycerin Bottle.

(*b*). *Section lifter.* Take a strip of tin or thin sheet copper, five or six inches long and three quarters of an inch wide, round the ends and bend it as shown in Fig. 10.

This little instrument will be found useful in lifting fragile sections from the saucer to the slip. When very fragile, the section must be floated from the lifter to the slide by putting a few drops of glycerin on the lifter just behind the section and tipping the former so that its point just touches the drop of glycerin on the slide.

FIG. 10. SECTION LIFTER.

(c). *Reflecting box.* Take a cigar box, seven or eight inches long, six inches wide and two or two and a half inches deep, and remove the top and one of the sides (leaving one side, the bottom and the two end pieces). Into this fit a piece of looking-glass, extending diagonally from the top on the closed side to the bottom on the open side. Finish the apparatus by covering with a plate of good clear glass. The apparatus is almost self-explanatory and will be found invaluable in making minute dissections, arranging specimens on the slide, and in fact any kind of work requiring reflected light. In use the box is placed so that the light, impinging on the surface of the mirror, is reflected up-

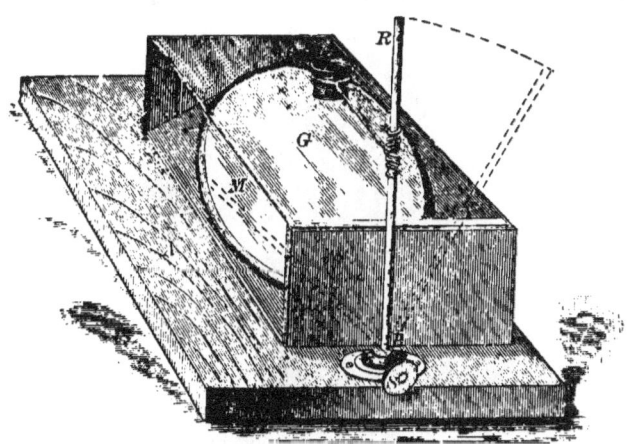

FIG. 11. REFLECTING BOX.

ward through the glass cover plate, thus enabling the operator to see the slightest defect or fold in the arrangement of his specimen. The box is greatly improved and enhanced in value by the following additional apparatus, which converts it into a true mounting microscope. Take a piece of wood eight inches wide, one foot long and one inch thick and in one end insert an upright iron or brass rod, one quarter of an inch in diameter and ten inches high. Twist a piece of stout copper wire so that at one end it will form a loop capable of receiving and holding a jeweler's eye-glass, and at the other end attach it to the upright rod in such a manner that it forms an arm four or five inches long which can be slid up or down upon the upright rod. The jeweler's eye-glasses or loupes are cheap and generally made of excellent lenses. I find it convenient to have four of them, of focal lengths ranging from a half inch to two inches, constantly at hand.

FIG. 12. WIRE CLAMP.

(*d*). *Clamps.* These may be bought ready made for 75cts. per dozen, or they can be fashioned out of good steel hair pins. Fig. 12 represents the best form of clamp for almost any kind of mounting work. It is self explanatory

(*e*). *Cover-glass holder.* Take a piece of brass or copper spiral spring wire, such as is used on bells, and wind it lightly

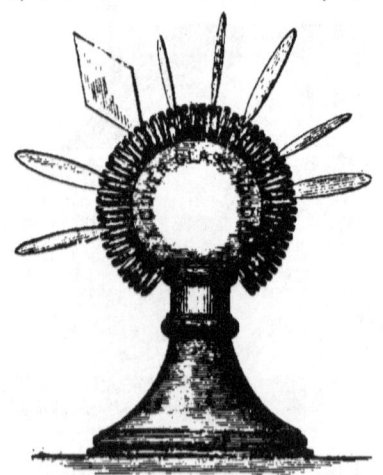

FIG. 13. COVER-GLASS HOLDER.

over a bit of cork or other round substance not over an inch in diameter. Fasten it to its place by sticking the ends into the cork and attach the latter to some heavy base which will keep it upright. The cover-glasses, rubbed and cleaned, are inserted between the spiral coils much as pen holders are inserted in the racks made of similar wire more loosely twisted. When the covers are inserted the whole should be protected by inverting a goblet over it.

(*f*). *Drying papers.* These should be made of the best Swedish filter or the best linen blotting paper, cut into a strips half inch wide by two or three inches long. They should be kept in a tightly covered box, protected from the dust.

The other instruments used in mounting have already been described in our earlier chapters. They consist of straight and curved needles, firmly set in good strong handles, camels-hair or sable pencils, watch glasses, etc. All of them should be clean and free from dust.

§ LXXVI. Arrangement of Work Table.—The work-table should be placed in front of a window giving a good light, and removed as much as possible from jar, dust, soot, etc. In St. Louis and in most large cities where soft coal is largely used, these precautions are more easily prescribed than followed. Rapidity and excellence of workmanship in mounting, as in all other mechanical operations, depend a good deal upon a natural aptitude for such work, but more on a proper and convenient arrangement of instruments and facilities, so that no time is lost in putting the hand directly upon the desired instrument or object. These, it is presupposed, are always kept in a proper condition—the knives, needles, scissors, etc., sharp and clean, the cover-glasses and slips thoroughly clean and conveniently arranged, the wash-bottle filled with distilled water, etc. Directly in front of the operator is his reflecting box, to the left of which and close by, is the bottle of cement. In front of the reflecting box, but placed so that it will not interfere with the light, is the vessel containing the objects to be mounted. To the right is the case containing the glass slips, ringed and ready for use, the cement hard and dry and the glass clean. Near by, in convenient reaching distance, are the glycerin bottle, the wash-bottle, a saucer of clean distilled water in which rests a camels-hair pencil, the little box of drying papers and other conveniences, among which I would especially mention a clean soft linen cloth for wiping slides, and a bit of chamois for giving them a final polish.

§ LXXVII. THE OBJECT TO BE MOUNTED having been removed from the fluid in which it was left after section-cutting, is placed in pure glycerin and allowed to remain there until it is permeated. It should be carefully examined with a Coddington lens or magnifying glass to see that it is clean and free from adherent embedding material and other extraneous matter. A glass slip is chosen, with a cell of proper diameter and depth and after being wiped with a moist cloth, is dried and polished with the chamois. Care must be taken that no particles of lint or fibre remain on the surface, either of the slip or cover glass.

The slip is now brought under the delivery tube of the glycerin bottle and receives a drop of glycerin, which should be placed in the center of the cell ring. In doing this it is well to expel a drop or two of glycerin from the point of the tube before allowing any of the fluid to touch the slip, since there is always more or less dust settling around the tube, and which is thus avoided. There is also less likelihood of air bubbles when this precaution is taken. Only a small amount of glycerin is needed, and beginners invariably put many times too much upon the slip.

The object to be mounted is now taken on the section lifter or with the forceps and deposited on the drop of glycerin on the slide; the latter is placed on the reflecting box and here the object is smoothed out and arranged exactly as it is desired it should remain in the permanent mount. If the object be one having one side or edge solid and the other ragged or broken (like a section of certain tumors), it should be placed so that the unbroken side is next the left end of the slip, for the following reason: when the cover glass is adapted the lower edge thereof is usually carried to the left edge of the ring and let fall toward the right. It thus has a tendency to straighten out and drive toward the right any filaments or parts of the mount that are easily moved.

§ LXXVIII. When the object is one consisting of several parts which must maintain certain definite positions toward each other, but which when in section are held together very slightly or not at all, other precautions must be taken in the preliminary arrangement of the slide. If celloidin has been used as an embedding material the described end is attained by leaving the section in the embedding mass and mounting it all together. The celloidin becomes entirely transparent when brought into contact with the glycerin, and does not interfere with the examination of the object with the highest powers. Under other circumstances the best device that I have yet found for fixing the objects in position consists in giving the slide a very thin coating of gelatin before placing the object upon it, and adding the mounting fluid only after the object has attached itself lightly to the slide. In such cases the cover-glass must be applied very soon after the glycerin is added or the latter will re-dissolve the gelatin and the work must be done over again.

Air bubbles are gotten rid of, by slightly warming the slide and puncturing each one with a hot needle. With a little practice, however, the glycerin may be delivered from the bottle without a particle of free air.

§ LXXIX CLOSING THE CELL.—Take the cover-glass by the edge between the thumb and forefinger of the right hand, and bring it down at an angle of 45° until the lower edge touches the drop of glycerin in the ring, then carry it (still held at the angle mentioned) horizontally toward the left until the lower edge reaches the center of the cell wall, when the hold should be suddenly relaxed and the cover-glass allowed to fall upon the surface of the glycerin. It will fall gently and if properly done will cause the glycerin to fill the cell smoothly and without a bubble. It can now be shoved gently home, expelling the surplus glycerin as it goes. If when it is in place an air bubble is found within the cell, do not waste time trying to get rid of it. Long experience has taught me that one saves time, temper and material by lifting off the cover-glass, rinsing and wiping it and going through the operation of replacing it *de novo*, after adding fresh glycerin to the mount.

§ LXXX. WASHING THE SLIDE.—When the cover is safely home apply the wire clamp (Fig. 12) to hold it to its place. The slide is now ready for washing, to free it from surplus glycerin. This may be done by a stream from the wash-bottle, directed so that the water will not be forced up under the cover; or better still, the glycerin may be washed off with a large camel's-hair pencil plentifully supplied with water. This must be done most thoroughly, for upon the absolute removal of the surplus glycerin depends in a great measure the success of the mount. When the last traces are removed the slide is drained and wiped with a clean cloth, care being taken not to touch the cover-glass or clip.

The slide (the clip remaining in its place) is now laid flat on the table or mounting box, and with the little drying papers, every particle of moisture is removed from around the edges of the cover-glass, the paper being passed lightly but firmly around the line of juncture between the cover-glass and cement. To make assurance doubly sure, the slide should be allowed to stand a few minutes before applying the cement which is to seal the cell.

FINISHING THE SLIDE.

XIV.

§LXXXI.—CAUTION AGAINST FRESHENING. Some workers are in the habit of freshening the surface of the cell wall with a small amount of cement before applying the cover glass, claiming thereby to make a stronger and better cell. The plan is not a good one, and for the following reasons:

1. By restoring the volatile matter or a portion thereof, to the dry cell wall, the mass of the latter is sensibly augmented (to be diminished again in drying), thereby producing the very condition of things which I have shown (§ LV) to be the prime cause of leakage and "creeping" in glycerin mounts.

2. If for any reason the attempt at adapting the cover glass is a failure (as for instance, in case of the imprisonment of air bubbles or foreign matter, the displacement of the object, etc.), and it becomes necessary to remove it, it will be more or less smeared and soiled by the cement. The brightness of the glycerin will be affected by the agent, and the object itself is not unfrequently spoiled in the same way.

3. It is altogether useless and unnecessary. It was based upon the idea that when cement is applied around the edges of the coverglass none, or only a small portion of it, passed under the rim and that consequently a joint thus made is essentially weak and unreliable. This is a mistake. The fresh cement softens not merely the hard and dry surface of the cell wall with which it is brought into immediate contact by the pencil of the operator, but the softening process extends, by capillarity, some distance under the coverglass. This is easily proven by attempting to remove the cover from a well-made glycerin mount. In many mounts, some of them fifteen years old, examined by me recently the coverglasses were so firmly adherent that in most instances they were shivered to pieces by the effort to raise them.

§ LXXXII.—CEMENTING THE CELL. The slide, washed and thoroughly dried, is now freed from the clip which has hitherto held the coverglass in place, and is put upon the turn-table (§ LVIII). The coverglass must be carefully and properly centered, so that its rim is equidistant from and concentric with the inner edge of the cell wall. It is held to its place by atmospheric pressure, but may be moved from side to side by pressure made upon the rim horizontally and in a line exactly parallel with the surface of the slip. If properly done, there is no danger of air bubbles or other accidents. When the centering is accomplished, examine around the edges of the coverglass to see if any glycerin or fluid has escaped. If any be found it must be removed with great care and thoroughness by the use of the absorbent slips.

§ LXXXIII. The pencil is now dipped into the cement, the turn-table set in rapid motion, and a ring is spun around the edges of the coverglass. For glycerin, gelatin and other aqueous media, the white zinc cement described in § LVI is the best that I have ever used, though asphalt, shellac, and gold size, each have their admirers and advocates.

The first ring should be a light one, and as soon as it has been applied the slip should be removed from the turn-table and put aside until the cement dries or sets. The slip should lie flat while this process is taking place, for the reason that, otherwise, the softened cement of the cell wall would have a tendency to 'run' and to collect at its lowest point, thus distorting and injuring the cell.

The object of this light ring is simply to soften the cell wall and thus allow the coverglass to make an absolutely accurate co-adaption and joint. It is my habit to give the slide a thorough rinsing with a camel's hair pencil dipped in clean water, and dry it carefully before proceeding to apply the second and subsequent coatings of cement. The object of this is to remove the last traces of glycerin that might possibly be lurking around.

§ LXXXIV. THE NUMBER OF LAYERS of cement necessary to make a good, durable mount will depend on the depth of the cell, the thickness of the coverglass and upon the quality and character of the cement used. In the great majority of instances where zinc cement is used, from three to four layers will be sufficient. I am in the habit of applying sufficient cement to entirely hide the angle made by the coverglass and the cell wall – in other words, of using sufficient cement to make the line from the upper edge of the coverglass to the surface of the slip, straight or nearly so. The cements, enamels or varnishes used for finishing (§ XCI et seq.) will give the cell the appearance shown in Fig. 14.

FIG. 14. SECTION OF CELL.

This figure also serves to illustrate the point made above in regard to the softening of the cell wall (§LXXXI, 3). The darker part represents the old and dried cement, while the unshaded portion shows the manner in which the fresh and softened cement embraces the edge of the coverglass. Figure 15 illustrates a method by which a very strong cell may be made, when for any especial reason extra strength and solidity is required.

FIG. 15. SECTION OF CELL.

The result is attained by using a coverglass somewhat larger than the cell wall, so that a portion of the former projects all around, as shown in the figure. In using this form of cell, great care must be exercised in the process of drying after the preliminary washing. The drying paper must be passed under the projecting edge again and again, until every particle of moisture is removed.

§ LXXXV. A RAPID METHOD OF MOUNTING IN GLYCERIN, and one which answers very well for temporary mounts, is as follows: The object is arranged on the slip with a very small quantity of glycerin, so that when the coverglass is applied the fluid will not reach quite to the edges of it. The cement is applied in the usual way and runs under the cover until it meets the mounting fluid. Slides can be prepared very rapidly in this manner, and when carefully executed the mount will last a good long time.

§ LXXXVI. BALSAM MOUNTS are closed or cemented with balsamic cements, by first coating the edges of the coverglass with an aqueous cement—one made with gelatin, gum arabic or arabicin and insoluble in the balsam and its solvents. When this intermediary coating is thoroughly dry and hard, they can be treated exactly as glycerin mounts and be finished off with zinc white or any of the balsamic cements.

§ LXXXVII. The Pencil used for cementing the cells should be of camel's hair and not too small. The great mistake with many workers is the use of little, finnicky, fine-pointed sable pencils. The amount of cement taken up should be sufficient to sweep well around the cell in one motion. Another point which must be guarded against, if you would obtain a good smooth finish, is allowing the pencil to remain too long in contact with the ring of cement after it is spun. The benzol used as a solvent in most of the cements evaporates from the surface almost instantly, and if the pencil be held in contact with the ring a moment too long, the exquisite glazed surface formed by the 'setting' of the gum is dragged and broken.

§ LXXXVIII. Lay the slide flat. Whatever cement be used for closing and finishing a cell, to get the best results the slide must be allowed to dry in a flat position. This is a most important point, and one, the neglect of which is, I am sure, the cause of a great many failures in making handsome and lasting mounts. The reasons for this precaution are given above (§ LXXXIII) and on this account I prefer for the permanent preservation of slides, cabinets in which they maintain a horizontal position.

§ LXXXIX. The Slide is now practically finished and if due attention has been paid to detail in its construction, it will last for a great many years—though just how long the best work will remain unaltered is an open question. No mounting medium with which we are acquainted will make a preparation that will last indefinitely and keep bright and fresh always. The chemical effects of light and heat; the physical alterations produced by variations of temperature (shrinkage and expansion), although exercised within exceedingly minute limits; molecular changes in the glass itself and in the cements,—all tend, sooner or later, to destroy the labor of the cunningest hand and the most consciencious of workmen. One thing, however, is certain; the better the materials used and the more careful the workmanship, the longer will be the period of usefulness of the slide. I say 'period of usefulness' because a slide or preparation that may appear good and fresh to the unaided eye may have long since become valueless as a microscopical specimen. This is particularly true of balsam mounts whose edges have been left unprotected by a cell finish. In process of time the balsam becomes red 'or orange yellow, and semi-opaque or it so thoroughly permeates the object that the more delicate portions of the latter are lost to microscopical vision. Sometimes after many years it becomes so thoroughly dry that it loses its cohesiveness, and separates from the slip or cover-glass. The balsam mount, however, has one great advantage over one made with glycerin—it can be cleaned and handled with much less care, the cover-glass being supported at all points, instead of around the edges alone.

§ XC. STRIPING AND VARNISHING.—Cells finished with asphalt or any dark, shining cement, need nothing more, and may now be labelled and placed in the cabinet for future study or reference. Those mounted with zinc cement, however, even when the latter is highly glazed, will easily become soiled and the cells *will* show finger or dirt marks. For this reason they should be covered, at least partially, with a colored varnish or cement, which serves a twofold purpose, viz: protection and ornamentation. Where a colorless or transparent covering is desired, resort may be had to any of the quick-drying varnishes used by artists, or to damar dissolved in benzol. As these, as they are found in commerce, are usually too thick to flow freely from the pencil they should be thinned down with benzol before using.

§ XCI. COLORED VARNISHES.—The striping or ornamenting varnishes that have succeeded best in my hands are those made by rubbing up a dry coloring-matter in the solution of damar in benzol used for making zinc cement. They are very brilliant in finish, dry rapidly, and being to a great extent homogeneous with the cement of the cell, they attach themselves so firmly thereto as to become virtually a part thereof. The colors are those used by waxworkers and come in little vials costing from five to fifteen cents each. There are a vast number of shades of these colors, the most useful and brilliant however, being the blues and reds, ultramarine for the former and scarlet or crimson for the latter. In preparing the varnishes follow the directions given in § LXI for making zinc cement, using the coloring matter instead of the zinc oxide.

§ XCII. ENAMEL CEMENTS.—I have given this name to a series of finishing cements and varnishes made by myself after the following method: Good boiled linseed oil is cut with sufficient benzol to make it thin enough to pass through filtering paper. After filtration the benzol is driven off by heating in a water bath, and a stream of oxygen gas is conducted through the residue, being continued so long as it will pass, the oil being rapidly oxidized and converted into a semi-solid, viscid mass, somewhat resembling india-rubber in appearance and many of its characteristics. This mass is soluble in pure benzol, making a varnish which dries rapidly and is of extraordinary tenacity and brilliancy. Into this I grind the colors exactly as in making the zinc cement before referred to. The "strippings" from painters' oil and paint buckets, dissolved in benzol, make a cement very similar to, if not identical with that made by oxidation of boiled oil. In this case the process of oxidation is accomplished by the atmospheric air, only more slowly than by the use of a stream of gas.

§ XCIII. The Pencil for striping should be much sma than that used for applying cement, and should come to a fine point. It should be passed through the stopper of the receptacle, and be kept in the fluid when not in use. This rule applies to all pencils used in cements. This aperture in the cork or stopper should be bored to receive the handle and the latter should fit tightly.

§ XCIV. Application of the Colors may be made after the cement of the cell has become thoroughly dry, or they may be used as soon as the surface has become glazed. The slip should be carefully centered upon the turn-table, as the least deviation will be doubled by rotation and the result will be marked eccentricity.

Some workers cover the whole cell-wall with a transparent varnish after application of the striping, but I do not find that this conduces to either beauty of finish or strength, and consequently the operation of striping finishes the slide, which is now ready to be labeled and put away.

End of Part I.

We have followed the object from the crude materials to the finished preparation, taking up each step in its progress and treating it in extenso and in so plain a manner that anyone may, by following directions, make as good a slide as those coming from Cole or Bourcogne. Part II will deal with special methods of work and research.

APPENDIX.

PARENCHYMATOUS EMBEDDING, AND DRY SECTION CUTTING.

In the ordinary practice of microscopy, whether for clinical purposes or for the study of tissues, it has hitherto been customary to cut quite a large number of sections, from which only the best were selected for use and the balance thrown away. The exigencies of modern biological study, however, are such that this method cannot be pursued, since every section of the material is required for examination and must be mounted in definite order with relation to the preceding and succeeding section. These sections must also be of an uniform and definite thickness throughout. The demands of the biologists were promptly met by the instrument makers, and any of the machine microtomes described in the foregoing chapters will cut the sections as required, providing that the material has been properly prepared and embedded.

PREPARATION AND EMBEDDING.—Various methods have been described in the journals of microscopy and the natural sciences for thus preparing material for the microtome, all of them modifications, to a greater or less extent, of the method first suggested by Professor Lunge of Heidelberg in 1874, and known as the PARAFFIN METHOD. The steps in this method embrace—1, dehydration, first with 95° and afterward with absolute alcohol; 2, removal of alcohol by chloroform, turpentine, or benzol; 3, infiltration with paraffin dissolved in one of the media last named; 4, infiltration with pure paraffin; 5, embedding.

1. *Dehydration.* If the object has been hardened in absolute alcohol this step will not usually be necessary; but if any of the aqueous hardening mixtures or solutions has been used, the object must be transferred to alcohol of 95° and there left for a length of time dependent upon its mass, the nature of the tissues, etc. From 95° alcohol it should be removed to absolute alcohol, in

both instances sufficient fluid being used to insure absolute dehydration. On this first step, and the thoroughness with which it is executed depends the success of each succeeding process.

2. *Removal of Alcohol.* The object is now removed into whatever fluid it is intended to use subsequently as a solvent of the paraffin. Oil of turpentine, chloroform and benzol have each been recommended at different times by competent authority. Turpentine offers the advantage of cheapness, but it is the least volatile, which is a grave defect. Probably chloroform is the most available. Whatever fluid be chosen, it should be abundant—not less than 15 to 20 times the volume of the object. In this the latter should be left for 24 hours, or long enough for the fluid to permeate the tissues thoroughly.

3 and 4. *Infiltration with paraffin.* This progresses in two stages or steps. In the first the paraffin is cut with from one-fourth to one-third of its volume with one of the above mentioned solvents. This should be done with the aid of a very gentle heat, applied cautiously. The temperature should never be allowed to go above 140° F., and if pure benzol be used 130° to 135° will be sufficient. The paraffin should be FFF, or the hardest grade. The temperature mentioned will keep it just at the point of fluidity. The object, removed from the liquid (No. 2) is placed in this and there left from 12 to 24 hours, or until thoroughly infiltrated, when it may be removed into pure paraffin of the grade above mentioned, heated just to the point of fluidity. Here it is to be kept for another day or day and night, when it will be thoroughly permeated with paraffin.

5. *Embedding.*—If a well-microtome be used, the object may be removed from the melted paraffin directly into the well of the instrument, arranged in position for cutting, as described in the chapter on section cutting (§ XXI, XXII and XXIII), and the well filled with the melted paraffin. In the more modern machine microtomes the well is, however, discarded, and a different procedure must be adopted, as follows: A box is made of strong paper, of a size sufficiently large and deep to receive the object and to leave from a quarter to a half inch of paraffin around and below it. This box is nearly or quite filled with the paraffin, and the object is arranged in it exactly as in the well of the microtome (see *ante*), and the whole is left to harden. This may be hastened by standing the box in cold water. When perfectly

solid and hard the embedded mass is hermetically sealed, and may be put away for an almost indefinite period, for future use, or it is ready to go to the holder of the microtome. Before transferring to this, however, the paper should be removed, and also enough of the paraffin trimmed away from around the object to leave it almost bare at the point of contact with the knife, and give the upper portion of the mass the appearance of a truncated pyramid.

SECTIONING.—This differs in no respect from the process as heretofore described. Those who have achieved the highest success in this line of work (Dr. Reeves, of Wheeling, W. Va., Dr. Simon H. Gage and others) say that the more direct and rapid the cut the better the results. This is also my own experience.

FIXING THE SECTIONS ON THE SLIDE.—In section-cutting, as hitherto described, the sections were removed from the knife to a common receptacle in which they were afterward freed from embedding material, stained, etc., and made ready for mounting. In dry section-cutting, especially for serial mounts, this plan is manifestly impracticable, and has been superseded by a very ingenious one of transferring each section, as it is cut, directly to the slide and fastening it in place by a transparent cement which, while resisting the cleansing and staining fluids used in subsequent manipulations, does not in any manner interfere with the optical and other qualities of the mount. This cement is simply a solution of pyroxylin or gun-cotton, or celloidin (which is a very pure pyroxylin) in alcohol and ether, and the solution mixed with clove oil, after the following formula, which is known as *Schallibaum's clove oil collodion:*

```
Pure pyroxylin, ............................ 2 parts
Sulphuric ether, ........                  15   "
Alcohol absolute............               10   "
Clove oil............                     100   "
```

A drop of this solution is placed on the centre of the slide and allowed to partially evaporate. The section is then removed directly from the knife to the slide, placed in position on the collodionized spot and gently pressed to place.

Prof. S. H. Gage, of Cornell, recommends *(Histological methods,* § 41, p. 25) instead of Schallibaum's clove oil collodion, the use of plain collodion, which he allows to dry on the slide and afterwards softens up with ether and alcohol when required for use.

I have no experience with this method, but from what I know of Prof. Gage's skill in technique, have no doubt but that it yields excellent results.

FREEING FROM PARAFFIN.—The slide is now transferred to a widemouthed jar containing xylol, benzol, or oil of turpentine. The former is preferable, as it is not so inflammable as the other fluids. In this it is left until freed from all traces of paraffin. This will be in from 20 minutes to half an hour, though the sections may be left in either of the fluids indefinitely without injury. When removed, after allowing the fluid to drain off as much as possible, the slides should be placed for a few moments in absolute alcohol to remove the last traces of the former bath, and then rinsed with pure, cold water. This must be done immediately, as the sections, unprotected by the coating of paraffin, dry and shrivel almost instantly. They are now ready for staining.

STAINING AND SUBSEQUENT MANIPULATION.—If the stain be a rapid one and in aqueous solution, it may be dropped directly upon the section, and the slip placed in a horizontal position until the tissue is colored. This can only be done, however, with those coloring mediums which act rapidly, or in ten or fifteen minutes. With the slower stains it is necessary to immerse the slide in a vessel containing the coloring matter, and leave it there until the tissues are sufficiently stained. This latter plan is the best under all circumstances.

The subsequent processes of dehydration, clearing, etc., are identical with those hitherto described in the body of this work, the only difference being that the whole slide is immersed in the fluids used for dehydration, instead of the sections alone.

LIST OF ILLUSTRATIONS.

FIG.		PAGE.
1.	Army Medical Museum Microtome	33
2.	Seiler's Microtome Attachment	35
3.	Bausch & Lomb Microtome (plate.)	36
4.	Walmsley's Microtome	37
5.	Section through Blade of Knife	38
6.	Section Knife	39
7.	Turn-table	68
8.	Griffith's Turn-table	68
9.	Glycerin Bottle	83
10.	Section Lifter	85
11.	Reflecting Box	85
12.	Wire Clamp	86
13.	Cover-glass Holder	86
14.	Section of Cell	93
15.	Section of Cell	93

INDEX.*

	PAGE.	PARAGRAPH.
Acid, Hardening Solutions	11	V
Acid, Osmic—See Osmic Acid		
Algae, preserving fluid for	11	V
Anilin Stains, Red	52	XLI
Blue	52	XLI
Green	52	XLI
Eosin	52	XLI
Fixing	56	(5)
Animal Tissues, Hardening	17	
Staining	46	
Apparatus for Mounting, General	8	II
" Staining	54	XLIII
Aqueous Mounting Media	83	LXXV
Application of Stains	54	XLIV
Colors	96	XCLV
Arabicin Cement	66	(i)
Areolated Tissue, Embedding	30	XXII
Asphalt Cement	62	(a)
Axiom, in Staining	54	XLV
in Cell-making	71	XLI
Balsam, Canada, see Canada Balsam		
Balsam, Hard, Mounting in	81	LXXXIII
Balsam Mounts, Finishing	82	LXXIV
Closing	93	LXXXVI
Balsamic Mounting Media	74	
Bausch & Lomb Microtome	36	
Bell's Cement	65	(f)
Bichromate of Potassium, Cleaning Fluid	60	LIII
Hardening Fluid	11	V
Staining with	53	
Bottle, Glycerin	84	(a)
Box, Reflecting	85	(c)
Camphor Water	77	(c)
Canada Balsam, Preparation of	74	(a)
Mounting in	80	LXXI
Mounting in, with cell	81	LXXII
Carmine, Ammonia	47	(1)
Beale's	47	(1)
Borax	47	(2)
Czokor's	47	(5)
Indigo	51	(2)
Picro	50	(1)
Thiersch's	48	(4)
Woodward's	47	(3)

*The Arabic Numerals refer to the pages; the Roman to the paragraph. Small numerals or letters in parenthesis refer to sub-sections.

INDEX.

Carmines, The, Simple Staining with	47	XXXVII
Fixing Solution for	48	
Multiple Staining with	57	(4)
Casein Cement	66	(h)
Cement, Arabicin	66	(i)
Asphalt	62	(a)
Bell's	65	(f)
Casein	66	(h)
Diamond	65	(g)
Gold-size	63	(b)
Marine Glue	65	(e)
Oxide of Zinc	64	(d)
Seiler's	66	(h)
Shellac	63	(c)
Cements, in General	62	
Qualities Desirable in	62	LVI
Enamel	95	XCIII
Cell, The	61	LV
Distortion of	61	
Mounting in Balsam with	81	LXXII
Closing the, in Aqueous Mounts	89	LXXIX
Cementing the, in Aqueous Mounts	92	LXXXII
Section of	93	LXXXIV
Celloidin, Embedding in	31	XXIII
N Y. College of Physicians' Method	32	
Cells, for Dry Mounts	71	LXIII
Of Lithage and Glycerin	72	LXIII
Rebaz's	72	LXIII
Tin-foil	72	LXIII
Wax	71	LXIII
For Aqueous Mounts	88	LXXVII
Cell-Wall, The	67	LVII
Dimensions of	69	LX
Importance of Seasoning	69	LXI
Chemicals, Required in Mounting	9	III
Hardening by	19	
Chloride of Gold, Hardening Solution	11	V
Staining	51	XL
Cleaning Slides and Cover-glasses	60	LIII
Cleanliness, Importance of	71	LXIII
Clamps, Wire, for Mounting	86	(d)
Cold, Hardening by	18	
Colophony, as Mounting Medium	75	(e)
Cover Glasses, Dimensons of	60	LIV
Cleansing	60	LIV
Holder for	86	(e)
Cutting Sections	40	XXIX
Damar, Gum, for Mounting	74	(b)
Colorless Solution of	75	
To Filter	74	
Decalcification	22	XIV
Deficiency of Balsam, to Remedy	81	LXXII
Desmids, Preserving Fluid for	11	V
Dehydration	97	(1)
Dessication, Hardening by	19	
Diamond Cement	65	(g)
Dry Sectioning (Appendix)	97	
Embedding	25	I

INDEX

Embedding Areolated Tissues	30	XXIII
Celloidin	31	XXIII
Choice of Material	26	XVII
Materials in General	26	XIX
Mixtures, Preparation of	28	XX
Mixtures, Methods of Using	28	XXI
In Machine Microtomes	31	XXIII
Processes in General	25	XVI
Parenchymatous	97	
Enamel Cements	95	XCII
Eosin, Staining Solution	53	XLI
In Multiple Staining	57	XLVIII
Fixing Solution, Anilin Stains	56	(5)
Carmines	48	XXXVIII
Fluid, Goadby's	11	V
Mueller's	11	V
Myer's	11	V
Wickersheimer's	11	V
Fluid, Mounting, see Mounting Media		
Preserving, Formulæ for	11	V
Freezing Mixtures	18	
Freshening, Caution Against	91	LXXXI
Glycerin, as a Mounting Medium	76	(a)
Jelly, as a Mounting Medium	77	(b)
Bottle	84	(a)
Rapid Method of Mounting in	93	V
Goadby's Fluid	11	V
Gold Chloride, for Staining	51	(3)
Solution of	11	V
Gold Size, Cement	63	(b)
Gum Damar, see Damar		
Hæmatoxylon, as a Stain	48	XXXIX
Aqueous Solution of	48	(1)—(2)
Alcoholic Solution of	49	(3)
Hardening, Agents	17	X
Processes	17	XI
Vegetable Tissues	18	XII
Heat, Hardening by	18	
Holder, Cover-glass	86	
Hydræ, Preserving Fluid for	11	V
Hones, for Section Knives	38	(e) XXVII
Indigo Carmine	42	XXXI
Infusoriæ, Preserving Fluid for	11	V
Instruments and Apparatus	8	II
Care of	42	XXXI
Knife, Section	38	XXVII
Shape of	39	
Honing	38	
Handle of	39	
Valentine's	35	
Larvæ, Preserving Fluid for	11	V
Liquidambar for Mounting	75	(c)
Logwood Stains	48	XXXIX
Simple Staining with	50	XLVII
Multiple Staining with	57	(3)
Marine Glue	65	(c)
Mastic, Mounting Medium	75	(d)

INDEX.

Machine Microtome,	Bauch & Lomb's	36	
	Description of	34	XXVI
	Embedding in	31	XXVII
	Seiler's	35	XXVI
	Walmsley's	37	
Media, Embedding, Choice of		26	XVII
Embedding, Materials Used for		27	XIX
Aqueous, Mounting in		84	LXXV
Mounting, see Mounting Media			
Meyer's Preserving Fluids		11	V
Microtomes, see Section Cutters			
Mixtures, Embedding, see Embedding			
Mount, Anatomy of		59	LI
A typical		12	VI
Balsam, Finishing		82	LXXIV
Mounting, Operations in		15	VII
In Aqueous Media		84	LXXV
In Canada Balsam		80	LXXI
Mounting Media		73	LXIV
Classes of		73	LXV
Resinous		74	
Aqueous		76	
Mounting in Aqueous		84	
Mueller's Preserving Fluids		11	V
Multiple Staining		56	XLVIII
with Carmine		57	(4)
Eosin		57	(2)
Logwood		57	(3)
Picrocarmine		57	(1)
Example of		58	XLIX
Nematodae, Preserving Fluid for		11	V
Nitrate of Silver, Staining Solution		51	XL
Osmic Acid, Hardening Solution		11	V
Solution of		20	
To Prepare Solution of		51	XL
Oxide of Zinc, Cement		64	(d)
Parenchymatons Embedding, Appendix		97	
Paraffin, Embedding in, Appendix		98	
Papers, Drying		87	(f)
Pencils, for Cements		94	LXXXVII
For Striping		96	XCIII
Preservation of Materials		9	IV
Preserving Fluids		10	IV VI
Picrocarmine, to Prepare		50	XL
Processes, in Preparation of Slide		7	I
Hardening		17	
Rapid Method of Mounting in Glycerin		93	LXXXV
Reflecting Box		85	(e)
Ringing Slides		92	LXXXIII
Section Cutter		33	XXIV
Army Medical		33	XXV
Bausch & Lomb's		36	
Bullock's		35	
Machine		34	XXVI
Seiler's Attachment		35	XXVI
Walmsley's		37	XXVII
Accessories for		39	XXVIII

INDEX.

Section Knife	38	XXVII
Knife to Sharpen	38	
Lifter to Make	84	(b)
Sections, Cutting	40	XXX
Staining, see Staining		
Serial, Cutting (Appendix)	99	
(Dry Appendix)	97	
Seiler's Cement	66	(h)
Microtome Attachment	85	XXVI
Shellac Cement	63	(c)
Sharpening Section Knives	40	XXVI
Silver, Nitrate, Solution for Staining	51	XL
Simple Staining, see Staining		
Slide, Definition of	59	LI
Washing the	89	LXXX
Finishing the	91	LXXXI
Slips, Glass, Cleaning	60	LIII
Dimensions of	59	LII
Preparation of	59	LI
Solutions, Hardening, see Hardening		
Staining, see Staining		
Cleansing	60	LIII
Fixing, for Carmines	48	XXXVIII
Anilins	56	(5)
Softening Animal Tissues	22	XIII
Spongy Tissues, Embedding	30	XXV
Staining, an Axiom in	54	XLV
And Stains	44	
Animal Tissues	44	XXXIII
Apparatus for	54	XLIII
Bibliography of	58	XLIX
Carpenter's Rules for	58	XLIX
Object of	45	XXXIV
Multiple	56	XLVIII
Simple	55	
In the Mass	54	XLIV
On the Slide (Appendix)	100	
Stains and Staining	44	
Anilin	50	XLI
Application of	53	XLIV
Dilution of	54	XLVI
Formulæ for, see Individual Stains		
General Table of	53	XLII
Selective	50	XL
Striping and Varnishing Cells	95	XC
Styrax, as a Mounting Medium	75	C
Table, Work, Arrangement of in Sectioning	40	XXIX
Arrangement of in Mounting	87	LXXVI
Table of Stains	53	XLII
Technology, Definition of	7	
Tissues, Hardening Animal	17	X
Hardening Vegetable	22	XIII
Decalcification of	23	XIV
Turn-Table, Type of	68	LVIII
Griffith's	68	LVIII
Method of Using	69	
Valentine's Knife	86	

INDEX.

Varnishing and Striping	95	XC
Varnishes, Colored	95	XCI
Vegetable Tissues, Hardening	22	XII
Softening	22	XIII
Walmsley's Microtome	37	XXVII
Washing Slide	38	LXXX
Wickersheimer's Fluid	11	V
Work Table, Arrangement of for Sectioning	40	XXIX
Arrangement of in Mounting	87	LXXVI
Zinc Oxide Cement	64	(d)

www.ingramcontent.com/pod-product-compliance
Lightning Source LLC
Chambersburg PA
CBHW020150170426
43199CB00010B/963